INTERDEPENDENCE

MEANING SYSTEMS

INTERDEPENDENCE

Biology and Beyond

KRITI SHARMA

Fordham University Press : New York 2015

Fordham University Press has no responsibility for the
persistence or accuracy of URLs for external or third-party
Internet websites referred to in this publication and does not
guarantee that any content on such websites is, or will
remain, accurate or appropriate.

Fordham University Press also publishes its books in a
variety of electronic formats. Some content that appears in
print may not be available in electronic books.

Visit us online at www.fordhampress.com.

Library of Congress Cataloging-in-Publication Data

Sharma, Kriti.
    Interdependence : biology and beyond / Kriti Sharma.
        pages cm. — (Meaning systems)
    Summary: "From biology to economics to information
theory, the theme of interdependence is in the air, framing
our experiences of all sorts of everyday phenomena. Indeed,
the network may be the ascendant metaphor of our time.
Yet precisely because the language of interdependence has
become so commonplace as to be almost banal, we miss
some of its most surprising and far-reaching implications"
— Provided by publisher.
    Includes bibliographical references and index.
    ISBN 978-0-8232-6552-7 (hardback) —
ISBN 978-0-8232-6553-4 (paper)
    1. Autonomy (Philosophy) 2. Biology—Philosophy.
I. Title.
    B808.67.S53 2015
    111—dc23
                                        2014045376

Printed in the United States of America

17 16 15   5 4 3 2 1

First edition

To Om and Jyoti Sharma,
generous and loving parents.

To Rulia Ram Sharma, Kanta Sharma,
Giridhari Lal Nagar, and Raksha Nagar,
cherished and respected grandparents.

And to all the ancestors
upon whom *this* life depends.

CONTENTS

INTERDEPENDENCE

INTRODUCTION

Taking Interdependence Seriously

You may have heard the phrase "It's all connected" before. What does it even *mean*? Does it mean that an ice floe calving off the coast of Antarctica is about to cause the electricity to go out in your neighborhood? Does it mean that an intangible something knits together and resonates through all living and nonliving things, and we can tune into its hum? Does it point to the global economy and the vast network of actors and factors upon which each of our lives and personal fortunes depend? Does it refer to deer eating grass and wolves eating deer, or to electrons influencing each other across a distance? Is it a scientific fact? A spiritual experience? A political reality? A marketing slogan? An ethical claim?

Of course, the phrase "It's all connected" has meant all of these things and more, depending on the context in which it is used. What is being emphasized here is that the theme of interdependence is in the air, framing our experiences of myriad everyday phenomena. "The network" may be the ascendant metaphor of our time. Yet, it is precisely because the language of interdependence has become so commonplace as to be almost banal that we may miss some of its most surprising and far-reaching implications.

Nowhere, perhaps, is interdependence evoked as vividly and often as in the biological sciences. The careful work of biologists over millennia has produced a massive body of observations that illustrate a vibrant, intricate, causally complex world, in which products depend on processes, processes depend on products, wholes depend on parts, parts depend on wholes, and living beings depend on one another for our lives. The widespread interest in and acknowledgment of the interrelatedness of natural phenomena in biology thus creates *the illusion* of a coherent view of interdependence.

Still, I contend, the ascendant view of interdependence at play in biology—as in popular culture—*is not a view of interdependence at all*. It remains a view of *independence*. By and large, we think that interdependence just means *"independent objects interacting."* We say that things interact strongly, weakly, reciprocally, sequentially, and so on, but their ultimate independence from one another remains intact. Indeed, "the network," which has enjoyed a place of privilege in biological thinking in the late twentieth and early twenty-first centuries, is just such a map of separate and interacting entities. As long as the ascendant view of interdependence continues to collapse implicitly to a view of independence, I believe that we continue to miss an important implication of our own findings. Our beautifully detailed observations of biological phenomena have been and continue to be interpreted as evidence that our world is composed of a great many independent entities coexisting in interaction. Instead, what these data may be suggesting is that our world is not composed of independent entities at all.

This is a book about how we might meet interdependence *as* interdependence instead of continuing to view our world implicitly as fundamentally or ultimately composed of *independent* entities. To take interdependence seriously in this way does not mean viewing phenomena as murky undifferentiated masses or as overwhelming tangles of connections. Though "interdependence" is often associated with "complexity" (in the sense of "made up of many things") and has therefore often been assumed to be an intrinsically "complex" topic (in the sense of "difficult" and "complicated"), what I hope to show here is that a consistent theory of interdependence can actually be refreshingly simple, useful, and clarifying.

If there is a transition being made—in biology as in physics, neuroscience, sociology, and more—from viewing phenomena as independent to viewing them as interdependent, we can conceive of this transition as happening in two shifts. The first is a shift from considering things in isolation to considering things in interaction. This is an important and nontrivial move; it is also a relatively popular and intuitive concept in those fields. To get to a thoroughgoing view of interdependence, I argue that a second shift is required: one from considering things in interaction to considering things as *mutually constituted*, that is, viewing things as existing at all only due to their dependence on other things. This second shift is potentially more subtle and difficult than the first, because though the first requires considering the mutual relations and influences *between* things, it does not actually require a change in the many habits and assumptions that usually commit us to viewing *things* as fundamentally independent. The second shift requires recognizing and addressing these very habits, which currently obscure a thoroughgoing view of interdependence.

This book makes these habits explicit at every step and offers alternatives.

It is an invitation to walk through the concept of interdependence patiently and thoroughly, instead of simply accepting it as a truism.

To illustrate the utility of a thoroughgoing view of interdependence in a particular case, I examine a specific biological phenomenon—signal transduction—*in the absence* of a variety of implicit assumptions of *independence*. Signal transduction is commonly defined as the conversion of signals from the environment outside of a cell into physical or chemical changes within the cell. Though signal transduction may seem like a highly technical and perhaps obscure field of study, it is a familiar concept in daily life, even if it is not commonly referred to by that name. Signal transduction is central to our usual interpretations of everyday events. It is commonly conceived to be the beginning of all sensory happenings. Your sensing of fresh-baked bread from a bakery down the street is said to begin when molecules released from the baking bread are bound by receptors in cells lining your nostrils. The receptors *transduce* this signal from the environment into chemical changes within the nose cells, which lead to chemical changes in cells within the brain and ultimately to the perception of a smell.

This conception of signal transduction derives from a view that in every moment, organisms are radically separate from an external world with which they interact. Again, this is the common contemporary usage of "interdependence," which first divides the world into things and then says, "Look at how closely they're interacting." I show how the standard model of signal transduction not only *depends* upon the concept of fundamentally independent objects but actually *reinforces* the very idea that objects are fundamentally independent. I then offer a view of signal transduction that does not rest upon the assumptions of independence posited and reinforced by the standard view. This is where the utility of a consistent, thoroughgoing consideration of interdependence comes into view. Such an account offers novel and useful reconsiderations of the relation of objects and subjects; of lower-level and higher-level phenomena; of agency and determinism; of stasis, change, and causal relations; and of the relation of physical and psychological phenomena. These are topics of central importance to biology, cognitive science, and indeed, to our everyday experiences of life and aliveness—hence, to "life as we know it."

Throughout this book, I ask, "What do phenomena depend on? What do objects depend on? What do organisms depend on? What does sensing depend on?" This is also a way of asking, "What does *your* world depend on? Your body? Your experiences? What does *your* life depend on?" By focusing on interdependence in biology, I hope that this discussion may become less of an abstract theoretical exercise and more of a personal affair. I make many references to "biology saying this" and "biologists doing that." I should be clear that

I also mean to implicate *you* in the term "biologist"—whoever you are, however you were educated, whatever you do to stay alive. A biologist, in this view, is not simply someone who knows certain scientific facts and conducts particular kinds of experiments. A biologist is someone who carefully observes life and living beings. We are each of us alive, and so we each already know a thing or two about being living beings and being related with other living beings. This book is also, then, an invitation to recognize oneself *as* a biologist and to deepen one's biological practice—that is, to learn about biology as it is produced through the practices of professional scientists, yes, but also to observe life and living beings carefully oneself and to be willing to be touched and changed by what one finds there.

In short, this book is intended for biologists, as defined earlier: undergraduate and graduate students; academic scholars who work in and across the fields of biology, philosophy, and cognitive science; and, really, anyone interested in sharpening and clarifying their views about interdependence. I have made an effort to write clearly and precisely, but without burdening the reader with technical or scholarly jargon in the main body of the text (discipline-specific terms tend to appear in the notes, for those more interested in the interventions this work makes in particular scientific and scholarly debates). I wrote this book both to generate good ideas—as good academic work does—and to make clear how those good ideas can influence our everyday lives. Indeed, this book was born primarily out of a conviction that a careful study of interdependence can change the quality of our day-to-day lives for the better. Here are, for example, some of the things one can find along this road:

We are not radically separate from what we commonly conceive as "external reality." We are always in touch with reality, and are completely at home in the world.

The world is not a place that is created once and then waits for us to discover it. The world comes into being moment by moment, dependent upon our participation.

This is why our being in the world—our participation in its making—is so central to its continued creation, and this is part of the goodness of living.

I have encountered much resistance to such ideas, both within myself and in others: "Yes, these are pretty thoughts. If only they were *true*." I call such instances of resistance "assumptions" throughout the book; I count and address sixteen of them within these pages. The purpose of carefully and systematically addressing these assumptions is to bring them to light so that they can be known, and then either consciously chosen or abandoned. Once certain assumptions are abandoned, interdependence may be experienced as obvious, clear—and, yes, even *true*.

A BRIEF SKETCH OF WHAT'S TO COME

The standard general model of signal transduction can be summarized in the following simple sentence: "The organism senses and responds to the environment."

The organism, the environment, and sensing as a link between them are categories that are central to the standard view. So is the patterning that relates all of these into a coherent account—that is, order itself. Instead of taking each of these as fundamentally independent categories *a priori*, I ask of each phenomenon, "What does this depend on?" Chapter 2 addresses the external environment or *objects*, Chapter 3 addresses *sensing and response*, Chapter 4 addresses the organism or the *subject*, and Chapter 5 addresses *order*. In Chapter 6, I highlight the works of a few of the many contemporary and historical thinkers who have taken interdependence quite seriously, summarize the book as a whole, and reflect on how new views of interdependence may shift our perspectives about the *kind* of world that we live in.

This is a short book that sketches a coherent line of thinking and introduces relevant topics without treating them exhaustively. As such, it is meant to encourage further explorations. My hope is that the reader will remain engaged enough to read the entire piece once through to follow the whole thrust of the argument and will remain curious enough thereafter to seek out the works of some of the wonderful thinkers who figure in what follows.

1

IT DEPENDS

Contingent Existence

SMALL WORLDS

Not existence, not reality, but algae: That is what I had entered graduate school in biology to study. I had wanted to research how cells organize themselves into multicellular collectives and what was necessary for the transition from solitary to multicellular life. I was enamored of a creature called *Gonium dispersum*,[1] a (barely) multicellular alga. The evolutionary ancestors of this genus are free-living single cells, and its evolutionary descendants are multicellular[2] (Figure 1.1). *Gonium dispersum* lives on the cusp of solitary and multicellular life. Most of the time, the alga exists as an eight-celled aggregate. However, sometimes, the cells separate from each other and go their separate ways—hence, *dispersum*. "So much like us," I thought. "The cells are one kind of living being on their own, and a whole new kind when they're all together."

These more or less multicellular organisms abound in the world. A sea sponge passed through a fine sieve will disaggregate into individual cells that will eventually find each other again and reform a whole, functioning sea sponge. The soil amoebae of the species *Dictyostelium discoideum* normally live as free-living single cells, but when they start to starve, they go multicellular. The cells signal to each other, aggregate, form a multicellular slug made up of hundreds of individual amoebae, and collectively migrate to a new location, forming a mushroom-like body, and then dispersing again as single cells (Figure 1.2).

These transitions from solitary to multicellular life (and sometimes back again) made me wonder how single cells even *recognize* each other. How do *Dictyostelium* cells find each other to form an aggregate? How do they sense each other? For that matter, how do they sense anything at all? Do they

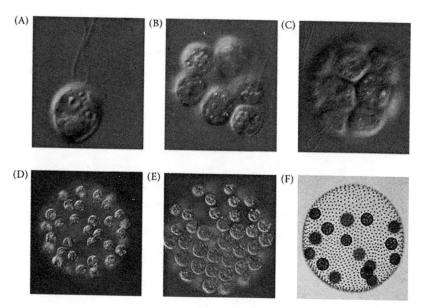

FIGURE 1.1. Algae of the family *Volvoaceae*, the genera of which are characterized by their number of cells, arrangement of cells, and differentiation of cells. The genus *Gonium* is characterized by four- to sixteen-cell aggregates, undifferentiated and arranged on a flat plane. (a) unicellular *Chlamydomonas reinhardtii*; (b) eight-celled *Gonium pectorale*, whose undifferentiated cells are arranged on a flat plane; (c) sixteen-celled *Pandorina morum*, whose undifferentiated cells are held together at the base and form a sphere; (d) thirty-two-celled *Eudorina elegans*, also undifferentiated and arranged as a sphere; (e) sixty-four-celled *Pleodorina californica*, whose cells are originally similar but differentiate and dedifferentiate over time as reproductive and nonreproductive types; (f) *Volvox carteri*, whose cells are fully differentiated early in development into reproductive and nonreproductive types. Courtesy of David Kirk.

differentiate between a fellow *Dictyostelium* amoeba and a sand particle, or bacterial prey? Are social organisms more responsive to members of their own species (called conspecifics) than solitary organisms? Do they have more faculties to perceive their conspecifics? For example, lions as social cats may be highly sensitive and responsive to the presence and behaviors of other lions, whereas tigers as solitary cats may be comparatively impoverished in terms of what they sense of other tigers. Similarly, cells that live in "societies" might have more "senses" (that is, more signal-response systems) attuned to cues from other cells (Figure 1.3, *a* and *b*), whereas free-living single cells might have relatively few means to sense and respond to other cells (*c* and *d*).

FIGURE 1.2. Development of the soil amoeba *Dictyostelium discoideum*. Under stressful conditions, the single cells aggregate and form complex multicellular bodies. Counterclockwise from bottom right: flat loose aggregate; domed tight aggregate; tipped aggregate; finger shape; hat shape; early culminant; later culminant (mushroom-shaped); fruiting body; fully differentiated spore body, with basal disk, stalk, and spores. The motile slug-shaped aggregate is shown on the bottom left. The aggregate will sometimes migrate to a new location in this slug-shaped form. Courtesy of M. J. Grimson and R. L. Blanton, Biological Sciences Electron Microscopy Laboratory, Texas Tech University.

I thought about "naïve" solitary cells, cells that have no means of perceiving their fellows. A conspecific could be brushing right up against their boundaries (*d*), and they would not be able to distinguish that from a ripple in the water. This is the same situation as if a cell were non-responsive to anything else in its environment. The cell could be totally surrounded by a certain sugar, but with no means of interacting with that sugar in any specific way (Figure 1.4*a*). The sugar *is* there to a human observer, but to the cell, it's *not* there. The cell does not sense it at all. If the cell lineage evolved proteins that bound the sugar, and this binding created some kind of response, then we could say that the cells became able to sense the sugar (Figure 1.4*b*). Now we might say that the sugar *is* there, not just to the human observer, but also to the cell. The cell, of course,

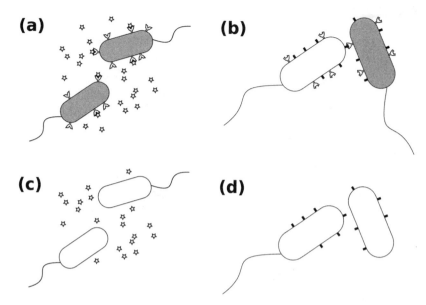

FIGURE 1.3. Cells and their "social senses." Cells secrete molecules into their environments (represented by stars) and have molecules upon their surfaces (represented by black rectangles). These molecules may be sensed by other cells, or conspecifics, as depicted in (a) and (b). Signal-response systems can bind the secreted molecules (as in (a)) or the cell-surface molecules (as in (b)) and produce a response in the cell (illustrated by the cell turning a darker color). "Naïve" cells that have no such signal-response systems, as depicted in (c) and (d), would be unresponsive to these molecules, and thus would be less capable (or perhaps incapable) of sensing the presence of conspecifics.

does not *call* what is there "sugar." However, I thought, "There *is* some *thing* in the medium, English-speaking humans call that thing 'sugar,' and the cell senses and responds to that same thing in some way."

I had started off asking, "Why do some kinds of cells respond to the presence of their fellows and others do not?" Then I asked the broader question, "How do cells evolve the ability to sense all kinds of things that their ancestors could not?" or in other words, "How do new senses evolve?" However, all of these questions seem to assume that *things are already out there*, and if organisms evolve the appropriate senses, organisms can start perceiving the things. Red objects have really been out there in the world for billions of years, and primates became capable of seeing that redness when they evolved eye proteins that are sensitive to red light.[3] Ultraviolet light was out there, and was

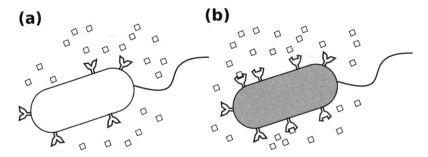

FIGURE I.4. Evolving novel signal-response systems. (a) A cell surrounded by sugar (white squares) is not necessarily responsive to the sugar in the absence of signal-response systems capable of binding the sugar. (b) Over evolutionary time, a descendant may arise with proteins that bind the sugar (transporters, signal transduction proteins, or other types).

perceived by bees and later detected by human ultraviolet-detecting technologies.[4] Similarly, sugar was out there and at some point in time began to be sensed by cells.

However, it is as reasonable to ask, "How do new things come into being for organisms?" as it is to ask, "How did new senses evolve?" In other words, when cells begin to sense sugars, a new "thing" comes into being for the cells. In a specific and important sense, the sugar did not exist as a *thing* before cells started to interact with it. Thus, organisms could be seen as "thing makers" in a world that is not necessarily full of things prior to the thing-making activities of organisms. I unpack the idea of a "thingless world" at length in Chapter 2 on objects and the external world, the idea of "thing-making activities" in Chapter 3 on signal transduction and sensing, and the idea of organisms as thing makers who make even *themselves* into things in the Chapter 4 on subjects and organismal selfhood.

The question "How do organisms sense the world?" is often read as, "How do organisms, with various degrees of completeness and accuracy, sense the one real world that exists independently of all of their perceptions?" Indeed, this was exactly how I read it when beginning to research cells and their senses. However, like many before me, I saw the problem in this formulation: There is no peeking at the world independent of perceptions. All we get as organisms ourselves *are* perceptions and conceptions. And so, at this point in the inquiry, I began searching for and contemplating ways to think about the real world that do not assume its existence independent of perceptions. That's when things started to get *really* interesting.

INTRODUCING KEY CONCEPTS:
REALITY, EXISTENCE, AND CONTINGENCY

First, we might ask, what would it even *mean* for the real world to exist independent of perceptions? To do this, we can begin by unpacking the term "existence." Philosopher Jay Garfield does this nicely:

> When we ask of a phenomenon, Does it exist?, we must always pay careful attention to the sense of the word "exist" that is at work. We might mean *exist inherently,* that is, in virtue of being a substance independent of its attributes, in virtue of having an essence, and so forth, or we might mean *exist conventionally,* that is to exist dependently, to be the conventional referent of a term, but not to have any independent existence.[5]

Throughout this book, I use the words "existence" and "reality" carefully, as Garfield suggests. I use the term "inherent existence" interchangeably with "intrinsic existence," "intrinsic reality," and "substantiality." I use "contingent existence" instead of Garfield's "conventional existence," for reasons I will explain.

"Inherent existence" means that a phenomenon has some kind of *essence*—something that makes it what it is, independent of anything else. What is essential is therefore what is fundamental. Some commonly posited essential (fundamental) categories include objects, matter, fundamental particles, energy, space-time, physical laws, moral laws, logical statements, God, selves, subjects, soul, mind, and consciousness. Essential categories are *independent, unitary,* and *continuous.* They exist without any dependence on anything *other than themselves* for existence. This includes their lack of dependence on *perceptions* or *thoughts,* which is why essentialism must entail a metaphysical (i.e., nonempirical) claim. That is, whatever the essence of a thing is, it cannot be confirmed by perceptions, precisely because it has to be independent of perceptions. It is what it is, no matter what you, I, or anyone else sense, think, or say about it.

Most of us wake up in the morning assuming the *intrinsic* existence of an external world. We tend to intuit that something out there exists that is thing-like or substance-like—unitary, continuous, and independent (that is, capable of being what it is separately from other phenomena, other objects, and/or from ourselves as external observers). We are sure that walls really are walls, that they are external to us, and that they are walls independent of us observing them and calling them walls. Though essentialism is articulated as a formal philosophical position, more important, it is an everyday or common-sense orientation to the world: a folk theory. In other words, this orientation does not necessarily have to be adopted by rational subjects as a formal intellectual theory—it is more often simply a way of being that precedes formal theory. Physicists who calmly explain the ultimate existence of all of physical reality (including themselves) as nothing but energy in various forms, for example, could still subscribe to essentialism as a folk theory. Insofar as we intuitively relate to the everyday world as substantial and separate from ourselves—instinctively perceiving, say, flowers, dogs, or dollar bills as being unitary and continuous independent of other objects and of ourselves as observers—we are folk essentialists. It seems that most everyone is a folk essentialist; the interesting question is why this might be the case. (From an evolutionary biological perspective, it is also interesting to ask whether most *organisms* are folk essentialists). I argue against essentialism as a formal philosophical position, but am more interested in its pervasiveness as a folk theory. I would consider *myself* a folk essentialist who is curious about why I am such and whether I could be otherwise.

As common as essentialism is, it does have viable alternatives. One can reject the idea of essences, it is important to note, *without* rejecting reality or existence. To say that things have no essences is not to say that they are non-existent. It is to say that no essences are required for things to exist. Again, to lack inherent existence means to lack essence. To say "a flower lacks inherent existence" means "a flower has no essence." Whatever might be pointed to and called "a flower" is constantly changing and not discretely unified. The labeling of a flower as a flower, based on some notion of "a flower pattern" or "flower-like organization," stays the same over time despite the changing of whatever is being labeled a flower. Thus, "the things themselves" are not what stay the same over time—words like "flower" are. The sense of a flower's continuity over time is a kind of experience, not an autonomous feature of an external world.

The sentence "Seeds turn into flowers, and flowers turn into soil" may be thought to describe a certain natural, biological process. In carefully watching seeds turn into flowers and flowers turn into soil, however, what is observed is a seamless phenomenon. At no moment does a seed turn into a flower. An observer, by some criteria, labels a phenomenon "seed" and at some later point labels a "different" (by some criteria) phenomenon "flower."[6] Terms are used to demarcate *seamless* phenomena for certain explanatory purposes.[7] Sentences that relate these terms are similarly useful for explanatory purposes. As philosopher Richard Rorty writes,

> Perhaps saying things is not always saying how things are. Perhaps saying *that* is itself not a case of saying how things are. . . . We have to drop the

notion of correspondence for sentences as well as for thoughts, and see sentences as connected with other sentences rather than with the world.[8]

This leads to what I will call "contingent existence" or "construct." In looking at a flower, for example, one might find that there is no inherently existing flower, but there *is* a "flower" (the term). Whatever "is" is not *intrinsically* a flower. It has no continuous, unitary, or independent existence.

This may seem an obvious or trivial point. No one expects a flower to be permanent—we know that it is always changing. However, precisely by seeing *the flower* as always changing, we see it as *one thing* that is changing, not as a number of different things. No one expects a flower to be unitary—we know that some people will define the flower as the bloom atop a stem, and others will define it as the whole plant down to its finest roots. However, we may have deep expectations around what is certainly *not* a flower—a butterfly landing on a petal, for example. No one expects a flower to be independent—it does not pop out of nowhere but appears at a nexus of seed, soil, sun, and water. However, once it has appeared, it is said to exist more or less independently of the factors that have given rise to it—for example, the flower can be removed from the soil or can survive in the dark. Moreover, the flower is said to exist independently of anyone's seeing it as a flower. These notions of dependence, independence, and interdependence are, I think, the most subtle and interesting in discussing the idea of intrinsic existence, and they lead to a discussion of "contingency" and "contingent existence."

The use of the word "contingent" instead of "conventional" (as Garfield uses in the preceding quotation) is important for several reasons. First, "conventional" is already a specific use of "contingent"—contingent upon conventions. However, "convention" can have the connotation (1) of something transitory, or somewhat arbitrary (like fashion or fads), and / or (2) of something that has been *decided* by a group of people, either implicitly or explicitly. I find these contemporary connotations of arbitrariness and willful choice somewhat misleading within the present discussion. The intricate history that gives rise to the *obviousness* of flowers—their visual sharpness and tactile solidity within human sensory systems, the history of a drive to be near them and to explain them, and the origin and continual use of the word "flower" itself, for example—suggests that the perceptions, recognitions, terminology, and explanatory values that gave rise to flowers can hardly be said to be *chosen* by any human individual, or even tacitly agreed upon by any human collective.

You and I never *chose*, in any meaningful sense, to recognize particular patterns as flowers; neither did we choose the word "flower." I may find myself experiencing something as vividly and obviously a flower—maybe a daisy, with white petals and a center as yellow as yolk. What explains this experience? I experience a sense of contrast with a background. Though this is obvious, it is also nontrivial. I do not experience the patterns on daisies that become visible under ultraviolet light. I have no word for that pattern. I do not experience the phenomenon of white spears of tissue emerging radially from a yellow circle as "grass," "coffee," or "Volkswagen"—nor could I choose to. Too much has already happened—I developed and was socialized throughout childhood, I have already had countless experiences that I am sure are *similar* to the present experience (never the *same*, but similar *enough*, in particular ways), and those experiences have been grouped together as "flower." I rarely or never experience myself as consciously choosing to do this grouping. No one around me has much of a choice in the matter; nor did the ancestors who first started to use the word "flower," and nor do *groups* of people experience themselves as coming together to decide through political consensus that, yes, we will experience this phenomenon with these boundaries and call that "flower." It is just that if everyone doesn't use "flower," it will simply stop working. There is thus nothing arbitrary about "flower," in the sense of it being somehow ahistorical or freely interchangeable with alternative terms, perceptions, conceptions, or experiences. In fact, I can think of little that is under as much pressure to stay the same.

I use the word "contingent" instead of "conventional," therefore, to move away from connotations of arbitrariness and choice. The dynamic, intricate, highly interdependent and highly ordered processes that give rise to the term "flower" over and over again are *precisely* what make flowers appear so obvious, vivid, and *stable* as objects. The dubious essentialist assumption is thus a relatively simple one: it is that obviousness, vividness, and stability are properties that somehow inhere in objects, instead of experiences that arise at the nexus of active living bodies, percepts, and values. An alternative approach would be to carefully detail the processes by which objects *come into being* as objects—that is, as continuous, discrete, vivid, and efficacious—contra the essentialist assumption that continuity, discreteness, and so forth are properties that somehow *inhere in* objects. Part of the work of this book is to give such an account.

FEATURES OF CONTINGENTISM

The notion that essential properties inhere in objects is often evoked to explain why phenomena that are not observed still *exist*. The intuition is that bacteria, deep sea creatures, distant stars, and, for that matter, ordinary objects like tables and chairs *exist* even if they're not observed or even never observed. However, none of these can exist as *things*—or even as formless substance or formless void—in the absence of observers. I walk through these problems in depth in Chapter 2. For now, suffice it to say again that the discussion of what the existence of unobserved phenomena means depends critically on how we understand existence.

Throughout this book, I use the term "contingentism" to denote a way of understanding existence that differs usefully from the standard essentialist view. These are some of the general features of contingentism:

1. Inherent existence means existing by virtue of having some essence.
2. Contingent existence means existing dependent on other contingently existing things.
3. Things exist contingently—dependent on other things, including, for example, observers that relate to them as things.
4. Critically, however, observers themselves exist contingently (dependent themselves on bodily organization, objects, and lives) and do not exist inherently.
5. If *everything* exists contingently, nothing exists inherently.
6. Thus, the thoroughgoing interdependence of things and their lack of inherent existence are simply two sides of the same coin.
7. Sameness and differences exist contingently and not inherently. Thus, things are not inherently the same as each other (monism), nor are they inherently different from each other (dualism).[9]
8. To say that "all things have no essence" is *not* to say that "all things have some essential sameness—that is, their nonessence." There is no need to posit some essential, self-presented "void" *behind* contingent existence.
9. Contingent existence is not a "lesser" kind of existence—"merely" contingent as opposed to "truly" substantial. Contingent existence *is* existence.
10. "Contingent existence" itself, as a contingently existent concept and term, itself has no inherent existence, and need not be reified. It is employed with reference to certain goals and values, like any concept, term, or thought system.

This is a dense network of concepts to outline. I will come back to them and flesh them out many times throughout this piece, instantiating them in real examples. Again, I am interested in using these views to help elucidate how biologists might conceive of what it means for organisms to sense and interact with an environment or external world, and how biologists might conceive of what it means for organisms and environments in general to depend upon one another. Once again, when I use the term "biologists," I mean to include *you*—a careful observer of life and living beings.

It is worth briefly discussing points 7 and 8 here. Contingentism assumes neither inherent sameness nor inherent difference. For example, no one would say that food is the same as excrement. The two are different—but not *inherently* so. We treat them differently. When asked *why* we treat them differently, we might claim that it is because they *are* different, or because they have different properties, as a way of justifying our differential treatment. From a contingentist perspective, there is no reason to talk about the properties possessed by food that make it *inherently* different from excrement. Differential treatment is quite justified enough *in the absence* of any inherent difference. In other words, there is no reason to search for some "real difference" between things in a way that suggests that the difference is somehow intrinsic to or *in* the things. A contingent difference—one that arises dependent on percepts, concepts, and practices—is quite sufficient to justify differential treatment. Meaningful differences, with consequences that matter for us, can be maintained without some notion of inherent differences. The criteria for designating food arise contingently and the criteria for designating excrement arise contingently. Any sameness attributed to food and excrement is contingent, as is any difference.

To summarize: There is no essential difference between things, but that does not mean that there is no *contingent* difference, nor does it mean that there is some essential sameness. This is particularly important to bear in mind lest we start to think that to say "all things have no essence" is to say "all things have some essential sameness, that is, their nonessence." Importantly, contingentism does not posit some substantial, inherently existing "nothingness" or "void" behind things as they appear. That would be nihilism, not contingentism. And nihilism is just one of the many things that contingentism is *not*.

WHAT CONTINGENTISM IS *NOT*

The idea that the world that organisms inhabit depends on their perceptions is easy enough to understand and may even seem obvious. For example, our experience of a colorful world obviously depends on our own visual systems. Yet the idea that the world is not strictly independent of perceptions may seem bizarre, foolish, and rash, or at the very least, exceedingly self-centered. We have, then, a claim that on one hand seems obvious and, on the other hand, seems outlandish—a recipe for interpretive disaster, or at least perplexity. We therefore need to proceed into this tricky terrain with caution, lest we think it so simple and banal that it has already been completely surveyed, or so difficult and bizarre that it is not worth surveying at all.

Before going any further, let me outline up front what contingentism is *not*. Contingentism is not a claim that the world does not exist. It is not a claim that organisms somehow conjure the world through their senses. It has little to do with "the law of attraction"[10] and other positions that articulate the interdependence of mind and matter as having to do with being able to alter (intrinsically existent) physical substance by altering (intrinsically existent) mental substance. Contingentism is not a claim that the world is an individual product of individual minds, forever separate and unable to share commonalities. It is not a claim that assertions of truth are impossible to substantiate, or that they are useless. And it is certainly not a call for an end to everyday actions, scientific practice, collective movements, reasoned and impassioned discussion, moral imperatives, or any other attempts we make to understand the world and act effectively in it.

Rather, contingentism is an attempt to understand reality in a way that *accounts for* the full interdependence of perceivers and perceived phenomena—which means, necessarily, not taking either of them to be intrinsically existent. From a contingentist perspective, it is not necessary to separate objects and subjects in order to make sense of and live well in the world. The object-subject divide, when reified, gives rise to all kinds of logical problems, and several practical problems besides. To remove it is to remove potentially unnecessary (and sometimes unhelpful) conceptual baggage. In other words, contingentism invites us to live without certain familiar but dubious assumptions, thus potentially enriching our lives as well as our views of the lives of other beings.

ENCOURAGEMENT TO STICK WITH A CHALLENGING TOPIC

In the course of writing this book, I have spoken with many people about existence. Most people, I have discovered, become somewhat uncomfortable when "existence" is the topic of conversation. The topic can seem challenging, not just in the sense of "difficult to apprehend" like a particularly complex and unfamiliar math problem, but also "difficult to take seriously" or "triggering an aversive reaction." One can speculate about various reasons why this may be the case:

A great deal seems at stake ("One minute you're having a conversation about new ways of conceiving existence, the next you're rambling on in public places about Time and God").

One does not want to be the kind of person who thinks or worries about existence ("Dressed in black and disaffected").

The topic seems already settled and irrelevant to everyday life ("To say the least. Whether or not we or anything else exists, we're obviously here and get on with our lives anyway, don't we?").

It feels overwhelming and complex ("I'm no good with theoretical matters").

One has heard all the same arguments before ("On one side are the people who believe the world simply exists, on the other are those who don't believe it. No one can occupy the latter position [or the former] and expect to be taken seriously").

Wherever the sense of challenge comes from, the point is that those of us who experience some form of discomfort when considering reality and existence are not alone. This category appears to encompass the vast majority of us. Fortunately, there are ways to turn discomfort to ease and gratification.

One way is through increased familiarity. To a certain extent, thinking in a certain way is a matter of habit. Ideas that are unfamiliar, difficult, or uncomfortable can become familiar, easy, and comfortable through repetition and use. This is the process of many kinds of learning, from acquiring a new language to, in this case, considering reality and existence in new ways. In addition to the development of familiarity, one can also address intellectual discomfort by recognizing and attending to its forms and sources. One of the cornerstones of contemporary formal education is "critical thinking." We are taught to think and to read such that we can apprehend, analyze, and evaluate what is being said. But careful thinking and reading also necessitate an attendance to the sensations that arise when one is faced with particular ideas— ideas that one considers "the same old," as well as those one considers outlandish, "the strange and new." Cognitive dissonance—the sense of discomfort that arises when one experiences something that is deeply contrary to prior expectations—is a well-studied psychological phenomenon from which none of us is immune.[11] By noticing that our bodies do indeed react in certain pleasant or unpleasant ways when we are faced with any given idea, we give ourselves the chance to distinguish what we like and dislike as well as the chance to go on engaging the idea. Critical thinking at its best, then, is actually *reflexive* thinking: the capacity to take into account one's own habits of thinking and feeling in the course of engaging ideas.

An example that psychologist James J. Gibson gives in his book *The Ecological Approach to Visual Perception* is useful here. Coming in from the cold and then touching a child's face, one doesn't react with alarm, inferring that the child has a fever. We are already in the habit of taking into account our own body and senses when making judgments, and we are capable of realizing that the experience of heat isn't produced by the child, but by the relation of one's cold hand and the child's skin. Similarly, by reflexive thinking, taking into

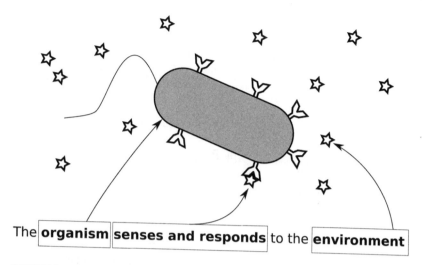

The **organism** **senses and responds** to the **environment**

FIGURE I.5.

account our own state and our own habits, mental and physical, we are better equipped to engage ideas about reality—the same old ideas, as well as strange new ideas.

SIGNAL TRANSDUCTION AND THE BOOK'S ORGANIZATION

The standard general model of signal transduction can be summarized in a simple sentence and schematic (Figure I.5).

The organism, the environment, and sensing as the link between them are categories that are central to the standard view of signal transduction. So is the patterning that relates all of these into a coherent account—that is, order itself. Though a standard view of signal transduction takes each of these categories to be intrinsically existent, the work of this book is to illustrate their *contingent* existence, the way in which they each arise dependent upon the others.

2

WHAT DO OBJECTS DEPEND ON?

Physical Substance, Matter, and the External World

In an obvious sense, this book is a real object in this world of objects that we inhabit. It is unified as one single thing with sharp boundaries and is clearly separable from a background of space. It is the same book as the book you were holding ten seconds ago. Moreover, if you needed any more convincing about its already vivid existence, you could point to it and ask anyone, "What is this?" They would likely confirm the obvious: "It's a book."

I have bored you with the mundane, perhaps. Fortunately, this mundane world of boring everyday objects is both exactly as boring as it seems and a little strange. What appears to be a world of self-presented and obvious objects can also be seen as a set of operations that we humans tend to perform: habits of assuming certain things about our world while denying others. In this chapter, I discuss six ways in which humans[1] tend to assume that we live in a world of intrinsically existent objects:

1. We assume the intrinsic boundedness and continuity of objects (for example, a book has clear boundaries and is the same book over time). However, boundedness and continuity are the experiences of observers, and not necessarily properties inherent to objects.

2. We assume the intrinsic boundedness and continuity of the fundamental parts of which objects are composed (for example, atoms). We might grant that objects themselves exist contingently, depending on our concepts of boundedness and continuity for what we take to be their "objectness," and yet still infer that fundamental particles of matter exist inherently. We might say that these particles are arranged in different ways and called objects by humans, but the particles themselves are substantial, indivisible, bounded, and continuous in a way that is independent of human cognitive processes. However, fundamental particles are inexpressible in the language of solid things or even in the language of fluid substance, except metaphorically.

3. We assume the intrinsic existence of something—causal power—that links one event to another. Even if we believe that there is no intrinsic boundedness or continuity to objects, or to the parts of objects, we observe that events occur in the world and infer that these events are produced by the power of things to affect other things. But our experience of regular occurrences does not point to the intrinsic existence of causal powers between intrinsically substantial things or of forces that inhere within, between, or upon things. In other words, causality is not a kind of glue that connects one event to the next.

4. We assume the intrinsic existence of (emergent) properties. For example, even if we do not assume that two hydrogen atoms and an oxygen atom exist intrinsically, we still think of water (molecules en masse) as being wet. In general, even if we accept that neither fundamental particles nor forces or energy exist intrinsically, we might think that everyday objects like tables and chairs still have properties, and that those properties emerge at higher levels of organization. However, properties themselves also depend on observation, and our experience of these properties refers to the close relationship of theory and observation, not to a quality intrinsic to objects.

5. We assume that the coordination of sense perceptions means that objects are inherently unified. In interacting with objects day in and day out, our senses sing in unison: "Here's a cookie! It can be seen, felt, smelled, and tasted." Moreover, people around us are constantly referring to these same objects, and seem to experience the vividness of objects with the same clarity and conviction as we do. But the unity and vividness of objects depends upon these forms of coordination between one's own senses and upon the coordinated actions of social members—unity and vividness are not *in* the objects.

6. We assume that even if "what is" does not exist intrinsically as things, there's still something there—it's just inexpressible. "Whatever it is," we might say, "no matter how subtle or formless, it surely doesn't depend on us." However, even our most subtle views of "what is" depend on our assumption of its necessary independence of us—an assumption that doesn't necessarily bear close scrutiny.

Again, "we assume" here does not mean that people are constantly making formal, logical propositions about the intrinsic existence of the external world. "Assumption" here means something closer to "intuitive, pretheoretical (and perhaps preconscious) habit" than to "product of conscious reasoned thought." Each of these assumptions, however, can be brought to conscious awareness and analyzed as if it were a formally articulated idea, which is what I will attempt here.

ASSUMPTION OF THE INTRINSIC BOUNDEDNESS
AND CONTINUITY OF OBJECTS

In one sense, objects do not appear simultaneously with the presence of observers, nor do they disappear in the absence of observers. Intuitively, their existence is independent of observers; your coffee cup does not disappear when no one is looking at it. In another and important sense, however, objects do disappear in the absence of observers. In the absence of a human observer, no one groups various elements of, say, a flower (petals, anthers, leaves) together and names them "flower."[2] In that absence, no one experiences the flower as a single continuous object—for example, calling the blooming flower and the withering flower the same flower—and no one picks out "a flower" from a background of space. For example, an observer such as an organism that could not distinguish light-bouncing-off-flower from light-bouncing-off-sky could not distinguish a foreground and a background. The idea that objects depend on the presence of observers is not solipsism (the view that the world exists only in one's mind); neither is it nihilism (the view that the world does not exist at all). There is a very precise sense in which objects depend on the presence of observers: only observers can perform the various actions necessary for experiencing phenomena as objects.

This is true even if the observer is not human and is "inside" the observed. In other words, the observer of the flower may be the flower itself (the object itself, the subject itself, the organism itself). The body itself can be organized into an object by an internal observer, who would need to do the same actions necessary for parsing reality into objects as an external human observer. The internal observer would need to group its various parts into one thing, locate itself and bound itself as separate from a surrounding space, and stitch its moment-to-moment existence together into a single continuous experience. This internal observer is commonly assumed to be human and called the subject or self. This will be discussed in Chapter 4.

To revisit the example of this particular book, we could ask, is this a single thing? It seems to have a unity—it hangs together. When you move part of it, the rest of it follows. However, when it is ripped in two along the spine, this one object becomes two separate objects. Is what we call one thing an aggregation of two smaller things, or for that matter, trillions of tiny things? This one, unified object is divisible into many tiny objects—atoms, and even further into fundamental particles like quarks. It is the observer who perceives a collection of things as a single thing, or conversely, who perceives a single thing as a collection of things. There is nothing in the paper that gives it some essential unity. Depending on preferences and goals, one could call it "one book" or "a million paper fibers."

Is this book the same book as it was ten seconds ago? Its particles are in constant motion. After a long time, it will become yellowed and worn. In thinking about a world of objects, we might think it obvious that objects are perpetually changing. It would be more surprising to us if objects stayed the same eternally. Very few people talk about objects as existing eternally, yet it is common to impose more subtle notions of permanence upon the world all the time. Though our commonsense conceptions of objects seem dynamic, they are actually subtly static. We accept that objects change, but in order to do so, we need to see them as the same thing changing. It is the observer with some concept of time who calls the past book and the present book the same book. There is nothing in the book that makes it the same book from one moment to the next.

This is the sense in which objects themselves exist contingently. To say that they do not exist intrinsically is to say that they do not exist as discrete, unitary, and continuously existing things independently of an observer experiencing them as such. It is not to say that books, flowers, dogs, and oceans somehow disappear into oblivion and reappear from nowhere, depending on whether or not someone is observing them. To speak of their existence as contingent simply calls attention to the kinds of processes that have to be performed in order for books and flowers to be perceived as if they existed intrinsically—processes that happen so consistently and habitually that they are usually glossed over in the course of ordinary daily life. In order for objects to appear as if they existed intrinsically, they have to be aggregated as single things, distinguished from background space, posited to be continuous in time, and radically separated from the perceiving subject. This is a lot of cognitive work that is often taken for granted.

ASSUMPTION OF THE INTRINSIC BOUNDEDNESS AND CONTINUITY OF PARTICLES

Even if we accept objects themselves to be dependent upon cognitive processes, we might think that something within objects makes them substantial "at bottom." Objects might be defined as specific arrangements of matter. We might say that though matter may be shifting and changing arrangement all the time, matter itself is substantial. This definition evokes the sense that objects are arrangements of some thing, some substance, usually expressed in terms of particles, atoms. This inference goes, "Even if the only thing that makes a piece of paper a piece of paper is the fact that we very easily identify it as one thing when it is really an aggregate of many things (atoms), these atoms themselves are not subject to such vagaries of convention."

The atomic theory attributed to the classical Greek philosopher Democritus describes atoms as very small, discrete, and extremely hard. Their hardness is what, in this view, makes atoms indivisible. Though what is now called the "atom" (literally, "uncuttable") as used in chemistry is a misnomer because what we call atoms are indeed divisible into electrons, protons, and neutrons and even further into quarks, fundamental particles are often conceived as something resembling Democritus's atom, something substantial and space-filling, like a ball. If we conceive of a particle as having spatial dimension at all, it must have separate parts—a top and a bottom, an inside and an outside. If it were solid all the way through like a billiard ball, even smaller particles would have to make up its filling. If it were hollow like a table tennis ball, it would be "filled" with empty space. However, "the particle itself"—the shell, the "outer layer"—would be divisible into smaller particles, themselves distinct from the empty space. A particle theory of matter—where particles are conceived as solid, space-occupying, separable from empty space, and indivisible—leads to a regress. The search for fundamental particles—conceived of as the "building blocks of matter," the "stuff" of the universe—can be a search for something indivisible. But nothing that occupies space can be indivisible. Space-occupying particles can always be decomposed—if not empirically, then rationally—into even smaller parts.

Perhaps we are no longer so influenced by Democritus, and no contemporary physicist conceives of the physical world as made fundamentally of solid, discrete, space-occupying, and indivisible particles. The particle model of matter is just a model and is useful for various purposes at the level of physics and chemistry; this does not mean that matter is particles. We could imagine this book as very tiny spheres (particles) held together in a particular arrangement by some kind of glue (forces). This model is very useful for helpful for predicting and controlling certain everyday experiences, even if we know that bulk matter is not actually spheres and glue.

A seemingly more sophisticated way of talking about particles might be in terms of forces. Even if they are divisible in principle, atoms tend to hang together over long time spans. They are indivisible for many intents and purposes, which is why we might think of them as "fundamental constituents of matter" or as "building blocks" of the world. The word "hard" or "solid" in the context of particles does not have to mean "physically tangible" in the way that Democritus described particles as much harder than any diamond. "Hard" can also mean "difficult to take apart." It is difficult to split an atom physically, even if we can decompose it into parts analytically. Tremendous force holds the parts of atoms together, such that at the level of chemistry, we can talk about things like elements and make predictions about how atoms will interact. (In the next section, I will talk more precisely about what is meant by "force," but for the purposes of this section, I use the word "force" naively.)

This piece of paper could be said to be composed of atoms, which are themselves 99.9999 percent empty space yet are also considered "space-filling" and thus in a sense possessing measurable volume. Moreover, they are considered to be massive—that is, as possessing mass. However, the space-filling property is simply the observed fact that atoms do not overlap. That is, even if two atoms are mostly empty space, the two empty spaces do not overlap. In the macroscopic world, two magnets of the same charge can be brought together only up to a certain point. No thing separates them, yet they repel one another, and so cannot exist any closer together. Analogously, atomic nuclei can only be brought so close together.[3] In the case of the macroscopic magnets, the boundaries of the magnets are seen clearly as their surfaces. However, if we put a collection of such magnets in a box, and if we noticed that they do not touch each other and that the distance between them is regular and measurable, we might consider their boundaries to actually extend beyond their visible surfaces to include a "field" of sorts, which is just air. This is in a crude sense how we might imagine atoms,[4] bounding them as spheres, though the sphere has no surface but is simply a way of describing how close two nuclei might be able to come to one another. In that sense, the space-filling property of atoms does not describe its own "actual volume" calculated from some bounded surface; rather, volume is relative, in that it is the relation between nuclei.

The point is simply that whatever we interact with in everyday life as objects, its "thingness" does not necessarily come from substantiality "at bottom." Empirically, physicists examining matter at the smallest levels don't generally find "stuff" in the way in which it is commonly conceived—that is, bounded, space-filling, and massive. Rationally, a theory of inherently existing individual particles is undermined by its internal contradictions. There cannot be such a thing as an indivisible particle if it has a dimension in space.

ASSUMPTION OF THE INTRINSIC EXISTENCE OF (EMERGENT) PROPERTIES

If pressed in this way, many will agree that matter is not in and of itself solid substantial "stuff." Yet, in general, we don't lose much sleep over this, and we consider it largely an academic curiosity. This is because the commonsense, everyday world of solid substantial stuff still appears, day in and day out. Stuff can and does emerge in everyday life in the absence of essential substance "at bottom."

Matter in aggregate is considered the purview of biology. Biology is about cells and biochemistry, organisms and ecology, matter on an aggregated and intricately arranged level. Why, then, have we focused so long on chemistry and physics in a discussion about biological interactions—that is, the interactions of organisms and objects? Since organisms interact with matter in aggregate, why consider particle physics and chemistry, which describe matter on a much smaller scale than experienced by organisms?

I focus on physics and chemistry because ideas about physics and chemistry are often in the background when we are considering objects, organisms interacting with objects, and organisms as objects. One common way to look for the inherent existence or substantiality of objects is to look for it "at bottom"— that is, to look for what it is within everyday objects that is inherently existent and thus can be aggregated to make the substantial things that organisms interact with. This is a reductionist approach, which was used in the previous sections. I have argued that using this approach, one can find no substantiality "at bottom."

Another approach is a holist approach—looking for substantiality "at the top." One can accept that matter is not substantial at the smallest levels but claim that at larger, everyday levels, substantiality arises through emergence. Emergent properties refer to the properties of the whole not found in parts. Water has the property of liquidity, but individual water molecules do not. Similarly, one might say that solidity is an emergent property of particles. Particles have no solidity, but particles in aggregate do.

First, what does it mean for a thing to "have" properties? An object cannot both be properties and possess properties. An object is none other than properties. We cannot even say, "An object is none other than its properties," because there is nothing independent of properties that possesses properties. In other words, there is no possessor of properties. Jay Garfield makes the point concrete. He writes, in the case of a tree, "Removing its properties leaves no core bearer behind. Searching for the tree that is independent of and which is the bearer of its parts, we come up empty."[5] That is, the whole is none other than the parts and none other than its properties. There is no intrinsically existing "ghostly level" called a "whole" that is in any way separate from the parts.

All this may sound a little strange given the popularity of the phrase, "The whole is greater than the sum of its parts." This is the intuition behind the idea of emergent properties. Indeed, the very idea of emergence suggests a kind of becoming, something appearing where it had not been present before. So when parts come together, it can seem like something novel emerges—hence the "greatness" of the whole relative to its parts. Whatever is happening in this shift from an assemblage of parts to a greater whole, it is not necessarily a shift within the thing. It is a shift in perception, when the observer begins to relate with a whole as a whole. A clay pot may have different properties from individual clay particles, such as structure, semipermeability and curvature, such that that it can hold liquids. To say that these properties emerge at a larger scale is a kind of metaphor with the connotation that the clay particles themselves, when aggregated in a particular way, acquire a novel state. What has happened, however, is that observers experience clay pots as having different qualities from clay particles just because they now relate to a clay pot as a whole. If the clay pot were not related with as a whole, no "emergent" property of structure, semipermeability, or curvature would be noted at all.

Properties do not inhere in objects; properties are themselves constituted by senses. They are whatever can be observed or measured. For example, a property of a copper wire is its electrical conductivity—a measurable quantity. We might say that the properties of copper atoms are what give a copper wire its properties, such as conductance. What we are doing, however, is evaluating our ability to make predictions about what is measurable (that is, observable) at a larger scale given a theory about smaller-scale parts. In other words, given a theory about copper atoms, we can predict that copper will conduct electricity. Of course, to say that larger-scale measurements (electricity) can be predicted based on theories about the smaller-scale parts (electrons) is a reductionist claim. Because this is common in the deployment of the language of "properties," I will address this reductionist claim first and then come back to the holist approach.

In saying, "the properties of the smaller-scale parts give the larger-scale whole its properties," we are describing a tight relationship between theory and measurement. Missing elements in the periodic table were predicted to have certain macroscopic (aggregate, emergent, measurable) properties based on a theory about their microscopic properties (their theorized atomic structure). That is, if we theorize that all the atoms that make up a metal wire have twenty-nine neutrons and twenty-nine protons, we will be able to make many predictions that will agree accurately with measurements we actually make on the metal wire. Often, it's the other way around: If we make a number of measurements on a metal wire, we theorize that the metal is made up of atoms with twenty-nine neutrons and twenty-nine protons. These tight relationships between theory and measurements are the result of a long and complex collective historical process: the creation and use of the periodic table of elements. We can explain the conductivity of a copper wire by appealing to atomic theory: The outermost electrons of copper atoms are somewhat unstable in that they can be more attracted to the nuclei of other atoms than to the atoms of which they are a "part," leading to the movement of electrons between atoms, which is expressed and quantified on a macroscopic level as "electricity." However, the word "explain" just means that we can describe one phenomenon in terms of another if we want to do so for certain purposes. We can use the language of electrons (smaller scale), and we can use the language of electricity (larger scale). Whenever elements predicted by atomic theory to have atoms with unstable outer electrons are subjected to an electric current, those substances conduct electricity. The shorthand for calling the conduction of electricity a unified, measurable quantity is "the property of conductivity." The language of "properties" describes a relationship between theory and measurement. It does not describe some essence that inheres in a thing. This view of properties will become particularly important in the next chapter, where we will examine how (smaller-scale) physical and chemical phenomena are said to give rise to (larger-scale) biological phenomena.

Another example of the relationship between theory and measurements is quantum electrodynamic theory (QED).[6] QED is an excellent theory that makes very accurate predictions that agree with measurements, yet makes no claim about the substantial properties of its referents. That is, using QED, one can make (probabilistic) predictions about how photons (en masse) will behave, but can make no claims about what a photon is. And this is precisely all one might want or need from a theory, for theories are tools—useful explanatory and predictive technologies, not ontological claims about the way "things intrinsically are."

"But," says someone with a furrowed brow, "isn't it uncanny that theories work? Isn't it amazing that the periodic table of elements as atomic theory existed before the discovery of many elements themselves? Surely this indicates that our theories are describing something independent of perceptions."

Let's say the periodic table predicted the existence of an object that has certain properties. We call whatever has those properties "copper." Now, copper is *by definition* whatever has these properties or measurable qualities. In other words, the theory itself defined the object class. That is, the phrase, "We discovered the object that was predicted by the theory," can often mean, "The set of observable properties that were predicted by the theory will be called a certain object."

In sum, neither reductionist nor holist views of properties help establish the intrinsic existence of things. A holist might say, "I don't care about the insubstantiality of things at bottom. Solidity occurs at macroscopic levels, where really big things like hands interact with really big things like apples. The same kind of thing happens when big things like proteins interact with big things like sugars. That's what I'm currently interested in." It is possible that there's no quarrel between this holist and a contingentist as long as (1) the holist does not use the language of emergent properties as meaning "properties that inhere in objects," (2) does not infer that the "big things" are intrinsically existing, and (3) acknowledges that things arise dependent upon cognitive processes (as described in the section on the inference of the intrinsic boundedness and continuity of objects). If the holist is asking not, "How do the properties intrinsic to objects emerge dependent on the properties of their constituent parts?" but rather "How do objects arise?" then he or she is asking the same thing as the contingentist. The language of "properties" does not add anything to the search for the inherent existence of things or help establish any such an existence.

ASSUMPTION OF THE INTRINSIC EXISTENCE OF CAUSAL POWERS

Even if one appreciates that there are no intrinsically existent things, no intrinsically existent constituents of things, and no intrinsically existent properties that inhere in things, we still live in world where things happen, regularly and often predictably. A moving billiard ball can cause a static billiard ball to move. We might accept that "billiard balls" are contingently existent. However, we know that one billiard ball can affect another. Billiard balls affect other billiard balls by causing them to move. They affect hands by getting in their way. They affect eyes by reflecting light. Importantly, billiard balls can be blown about by strong winds and crash into other billiard balls or tree trunks or anything else, and all of this can happen in the absence of any observer. Our sense of the intrinsic existence of objects (and of an objective world) may derive, then, from what we intuit as intrinsically existent causal powers.

In order to examine this intuition, we have to look at what we mean when we say that something causes something else to happen. This is no small task—the question of causation has occupied a central place in philosophy for centuries, and the definition of precisely what is meant by "this causes that" is still in the throes of lively debate today. What is important for the purposes of the present discussion is an examination of the common intuition that our experience of events as following one another regularly means, or is evidence, that there are inherently existent causal powers in the universe that *make* things happen regularly.

Philosopher David Hume noted that what we call "causes" begin as regularities, or our experience that events follow other events predictably.[7] Hume begins his theory of causation with the observation that human beings tend to cognitively connect phenomena. When one thing happens (we see a light switch being turned on), immediately, what comes to mind is another thing (the thought of a light bulb illuminating). Similarly, when a light switch is flipped, and a light bulb is illuminated immediately thereafter, we naturally connect the two phenomena. Thus, the attribution of connections between phenomena as inhering intrinsically within or between the phenomena, as opposed to within thoughts and experiences, is what Hume calls a kind of "transfer": "We transfer the determination of the thought to external objects, and suppose any real intelligible connexion betwixt them; that being a quality, which can only belong to the mind that considers them."[8]

Consider the billiard balls again. What is observed is that one billiard ball moves, it strikes another, and the second billiard ball moves. The thing that we call "cause" or "causal power" is never observed. A cause is not a thing (substance, "glue," "chain," or so forth) that connects two phenomena.[9] Nor is a cause a force that moves things, changes things, or holds things together. A cause is never itself available to observation, but is rather an inference that is drawn after making repeated observations.

Even the concepts of "energy" or "force" as used in physics cannot help the search for intrinsically existent causes, for these terms do not refer to any observable thing, no matter how subtle. "Energy" is a mathematical quantity that stays the same before and after other changes in a system, and is used to predict how systems change. (The concept of energy will be analyzed closely in the following chapter.) Force is a mathematical quantity that is used to help explain and predict whether things will move or not move, or whether they will be found near or far apart. For example, in the discussion of atoms above, I noted that the "hardness" of atoms needn't refer to solidity—"hard" simply means "difficult to take apart." Similarly, "force" in some contexts seems to mean "what keeps things together" (or perhaps, "what keeps things apart"), but more precisely means "what helps predict whether things will be observed together." "Atoms are held to each other by electromagnetic force, which is a 'medium-strong' force" means, "Most of the time, we observe atoms with other atoms, and we can predict the conditions under which they will cease to be observed together."

What we experience, observe, and narrate as the regularity and predictability of the universe is just that. Regularity and predictability themselves obviously depend on observers who group phenomena into patterns and relate them with other patterns. This "grouping and relating" process is analogous to how objects depend on observers who experience certain phenomena as bounded and continuous. Reference to anything—observable or unobservable, substance or force, power or law—that exists metaphysically beyond this regularity and predictability, in Hume's view, adds nothing useful or satisfying to the story. When this happens, that happens, but there is nothing intrinsically in or about "this" that makes "that" happen, nor is there anything intrinsically between "this" and "that" which connects them. In other words, there is nothing beyond regularities that explains them.

As an empiricist, Hume sought in his work to remove metaphysical categories from philosophical and scientific thinking and to promote a thoroughly naturalistic and empirical view of phenomena. Though his theory of causation is still widely considered the most thorough and influential work on the topic, it has also been criticized as austere, deflationary, and unsatisfying.[10] This type of critique also sometimes appears in reference to pragmatism: "Sure, some theories, beliefs, or views may be more useful than others, but are theories merely useful, or are they really true?" Similarly, with reference to causation, it may feel natural to ask, "Is the relation of one phenomenon with another a mere regularity, or is it a true cause?"[11]

If we want to distinguish between "mere regularities" and "true causes" simply in order to distinguish explanations that "break down easily" or have little predictive power from those that tend to hold under a wider range of circumstances, then this distinction is a practical matter that is carried out via everyday scientific practice and reasoning. For example, statements like "The sun rose because it's 6:00 a.m." are distinguishable from statements like "The temperature is rising because the sun has risen" by virtue of the very criteria we already have for what makes a good causal account: Causes are prior to effects, they describe a relation that holds in many circumstances, and so forth. These practices and criteria are being created, refined, and debated to this day, as is any other set of standards.

However, we may get the sense that, beyond the work of developing predictive technologies, the work of distinguishing "mere regularity" from "true causes" helps us to get at the intrinsically existent causal structure of the universe. The intuition here is that in order for causes to be really real, they must correspond to or reflect this intrinsically existent causal power or causal structure. In other words, the intuition is that there must be something other than regularities, that whatever corresponds to this "other than" is a real cause, and whatever does not correspond to the "other than" is a mere regularity. Yet, again, causes are never observed. What is observed is that when one event happens, another event happens, or, in the absence of one event, another event never happens. In other words, what are observed are regularities. So whatever is meant by a true cause as opposed to a mere regularity, it must be some nonempirical and metaphysical category.

For one who is used to seeing science and the discernment of genuine causes to be a process of uncovering the intrinsically existent causal structure of the universe, this view can feel unsatisfying and maybe even downright depressing. What appears to be implied is pessimism about knowledge: We will never know the real causal structure of the universe, and we will have to be satisfied with whatever observable regularities our limited and often faulty sensory and cognitive capacities deliver. However, such pessimism is hardly necessary or warranted. Indeed, one's reaction may be quite the opposite, for what is being suggested in this supposedly austere and deflationary view is a kind of optimism about our intimacy with phenomena.

For what if we have never been separate from the universe as it is? We are right here, in a world, a universe, and a reality that is very much immanent, accessible, and often predictable. The universe is predictable because we are here in it and have engaged from our birth as a species in developing and communicating predictive technologies. The causal structure of the universe is not metaphysically existent and inaccessible to us, a kind of crystalline net that always stands behind the veil of the patterns we experience, predict, and infer. Neither is the remarkable inviolability of laws such as the first law of thermodynamics remarkable because humans have now through tremendous effort and power broken through the veil to forcibly grasp and reveal something of this transcendent causal structure. Rather, what makes laws like the first law of thermodynamics remarkable is precisely the effort and ingenuity of the workers who parsed, related, and described phenomena in such a way as to render some statement inviolable. Such a process does not describe a prior separation from and subsequent revelation of the structure of the universe as it is. Such a process is better described as full participation in a universe that is rendered explicable, predictable, and understandable by virtue of the participation. It is our very intimacy with and participation in "what is" that gives rise to this very real, regular, and predictable world.

This is where contingentism offers a view that is genial, satisfying, and rich, contra the intuition that views that reject the intrinsic existence of metaphysical categories (including metaphysically existent causes) are austere, distastefully utilitarian, and pessimistic about the possibility of genuine knowledge. The sense of austerity derives from a feeling that nonmetaphysically based views deprive us from what we desperately need, leaving us somehow impoverished. However, on the contrary, such views invite us to see whatever we have immediate access to—the experiences that in fact arise for us and the knowledge that we in fact create and benefit from—as quite enough, quite satisfying, truly abundant, and endlessly developing. We will pick this matter up again in Chapter 5.

This is a book about interdependence, and naturally, causation is one of its central themes. This thread will be continued in the following chapters as we address a number of concepts closely related (obviously or nonobviously) with causation: energy and forces, change and stasis, agency and determinism, reductionism and holism, cognitive dissonance and consonance, natural order and laws, the experience of surprise, and predictive technologies. To make this thread easier to follow, I offer a contingentist view of causation in summary here:

From a contingentist perspective, even if we have

1. no intrinsically existent objects (this chapter, "Assumption of the Intrinsic Existence of Objects"),
2. no intrinsically existent constituents of objects (this chapter, "Assumption of the Intrinsic Existence of Particles"),
3. no properties inhering in objects, including causal power, or the property of being able to bring about changes in other objects (this chapter, "Assumption of the Intrinsic Existence of Emergent Properties"),
4. no properties within parts that give rise to properties of wholes (ibid., and Chapter 3, "Relating Physical and Psychological Phenomena"),
5. no causal links between objects (this chapter, this section),
6. no transfers of energy or force between objects (Chapter 3, "Assumption of Energy as a Kind of Substance"),
7. no uncaused agents (Chapter 3, "Relating Physical and Psychological Phenomena" and "Reviewing Sensing"),
8. no first, primordial, or original cause (Chapter 5, "Assumption of a Single Origin and a Linear History"), and
9. no intrinsically existent natural order or causal structure (Chapter 5),

still, an ordered, predictable, and explicable universe arises. Indeed, to state the case more strongly, the ordered, predictable, and explicable universe arises only because all of the aforementioned phenomena arise contingently, and do not exist intrinsically.

ASSUMPTION OF THE UNIFIED OBJECT OF SENSE PERCEPTIONS

Finally, perhaps the most powerful way in which the intrinsic existence of substantiality appears to be confirmed is through our immediate experience of objects as being obvious to many senses: "Of course it's there! If I can't see it, I can smell it. If I can't smell it, I can touch it." But this coordination of various sense perceptions doesn't point to the intrinsic existence of an external object. What it points to, rather, is that sense perceptions arise in a regular way.

One could say that when the senses change, the world changes. Someone with smudges on their contact lenses might see a blurry world. But "blurry" represents a contrast between the world when the observer is wearing smudged contact lenses and when she is wearing clean contact lenses. If she had been born with some "smudge" in the lens of her actual eye and there was no medical "corrective," she would have nothing in her experience by which she could compare a blurry world and a nonblurry world. There is no "taking off the lenses" and peeking at the "real world" behind them—this is how the world is to her. Such a person might tell others, "I see the world in a blurry way" or "My vision is impaired," but only because that was how she learned to describe her particular experience. Others would have communicated to her that her vision is atypical ("Oh, you can't see what's written on the blackboard?"). They might use the word "blurry" to describe what they have inferred about her experience from her behavior, from the studies they have done on her eyes, and from their own experiences of temporarily blurred vision. But "blurry" is typically used as a contrast to the world as seen by a majority who, through constant communication, create norms about how the world really is, which is sharp-edged.

Even though seeing the crisp edges of a book is an individual experience, the experience of clarity—that is, the sense that one is seeing the world as it really is—is, in an important sense, collectively created. People who have been told by a doctor that their vision is normal, and / or whose communications about what they see generally match up with the communications of the majority of other people's, will move through the world knowing that their vision is reliable and assuming that they generally see things as they really are. Conversely, they will know that their vision is reliable, and will therefore know that the things they are seeing are really there and they are seeing them as they are.

If we imagine a world where everyone, or even the vast majority of people, have smudged lenses, no smudge is actually perceived or conceived. Actually, there is no point in calling their lenses smudged—the lenses would be considered normal, in their natural state. The world simply is a soft-edged place, and most denizens of this world will be assured that they are seeing the world as it is. Indeed, this supposedly very different world is quite like the world we already inhabit, for our sensory capacities simply are what they are, and not what could be thought of as "optimal" in the sense of optimally suited for delivering to us an accurate picture of an autonomous reality. What our senses deliver to us simply is, and through coordination with social members, it becomes the real, shared world. This coordination is what biologist Humberto Maturana and cognitive scientist Francisco Varela called "bringing forth of a world through the process of living itself,"[12] and a coherent and careful exposition of this process has already been detailed in their books *The Tree of Knowledge* and *Autopoiesis and Cognition*.

In this "smudged lens" world, however, wouldn't the other senses besides vision deliver a worldview that seems to contradict the sense that the world is a soft-edged place? A soft edge could feel sharp to the touch, so the world wouldn't be soft-edged in every sense. However, in that world—again, as in ours—people would not necessarily actually sense a discrepancy between the visual and tactile senses. They would not say, "The edges of this book look fuzzy, but they feel sharp." There would be a certain pattern experienced between seeing and touching. Two things that look similar would also in many cases feel similar. This is how a feeling of coordination between the senses would be produced. How do any of us know that a book feels like it looks? How is something that looks like a clean edge "supposed" to feel? In nearly every instance of seeing a hard edge, when we've gone to touch that edge, we've felt a similar kind of sensation. We call both the visual sensation and the tactile sensation "hard-edged" because the vast majority of the time, they go together. If someone were to ask, "Does the book feel the same as it looks?" we would be a little puzzled by the question and say, "Yes—the book looks and feels like a book." The situation in this other, "smudge-lensed" world would be no different—to the inhabitants of that world, the book would "feel the same as it looks."

To use another example, in a world where people could feel differences between wavelengths of light, they could have a tactile sense of color. Every green object would feel similar in a certain way. If asked whether a leaf feels the same as a lime, they might say, "Yes, in some ways." Objects would both look and feel "green."

These examples help illustrate what it means for something to be real: The real is what arises for verbally, socially, and materially interacting human beings with more or less shared sensory capacities. Consider how often, day in and day out over the course of years, our ostensibly private and personal experiences are checked and shepherded, with the help of social members, into the appropriate category—real or not real:

"It was just a dream."
"Can you see what's written on the blackboard?"
"Are you tripping?"
"Did you ever see it happen again?"
"Can you explain how it happened?"
"Was that experimental result replicable?"
"It must have been a trick of the light."

Vast swaths of experience are placed in the category of "not real." Whatever is left over is the real.

This is not a clarion call to throw off the tyranny of social constraints upon reality. There are many good reasons to maintain these categories and to distinguish phenomena properly into one category or another (where, again, "properly" is socially defined). It matters for one's health that, seeing a fire in waking life, one is able to relate to the fire appropriately—that is, not to touch it and therefore not to get burned. Conventionally, one does this by placing the experience of fire in the category "real." That is, you don't relate to the fire as if it were a dream or hallucination that is therefore not capable of burning. You give it a wide berth. You come to know that the fire is part of the really real—that is, part of a world shared among social members—because you see others relating to it in the same way. Of course, having learned to trust our senses thanks to the validation of other people with whom we interact socially and materially, we usually tend to believe that whatever arises in our experience is real. Therefore, even if you're on your own, you will relate to the fire by giving it space and will assume that if another person were there, they would relate to it the same way.

But what if you're dreaming, or under the influence of hallucinatory drugs, or diagnosed with dementia? You may experience yourself as seeing a fire, or even as burning in fire. That fire is individually real, but it is not intersubjectively ("really") real: It does not have effects that are accessible to other people (your skin does not look burned to another observer, for example). If it is at all possible for this to be communicated back to you ("You're not burning, I see no fire, your skin is not burned"), you might therefore place the experience of fire and of burning in the category "not real"—or, if you're feeling charitable toward yourself, "real, but only (merely) for me." From the perspective of your experience of a burn, of course, that pain is real enough and is a valid object of sympathy. Yet, that experience may no longer count as real *even in your own mind*—even if you are experiencing it painfully, vividly—because you also know that what it means for something to be really real is for it to be what reliably arises for most people under particular physical and physiological conditions (wakefulness, sobriety, more or less "normal" physical and mental health, and so forth). You know that in your current state (drugged, drunk, half-asleep, and so forth), you don't meet those criteria, and so you experience yourself as burning—but not *really*.[13]

In sum, knowledge validation is a set of intricate processes—none of which points to the existence of inherently existent objects, but all of which point to how objects come into being dependent upon these processes of sensory and social coordination. One such process is coordination among various bodily movements and sense perceptions, which greatly enhances the vividness of objects and increases our conviction of their self-presentedness. Another process is coordination between social members, where continuously enacted and interacted perceptual consensus mark which of the many and diverse experiences that arise throughout our lives count as "the real world."[14]

ASSUMPTION OF NONIMPINGEMENT

Though one may agree that objects arise dependent upon "thing-ifying" processes, the following intuition can still powerfully remain: "Okay, so what is doesn't exist intrinsically as things. No big deal. There's still something there—it's just inexpressible. It is what is before anyone's perceiving, knowing, or thinking. Even if meteors, bacteria, and the sun don't exist as things independent of the experience of them as things, they surely still exist independent of you or me or any human being. The sun has burned long before we got here, and will likely continue long after we pass on. Something is there, even if it's inexpressible. To deny this is tantamount to believing that all that is depends on the experiences of the self or at least of human beings, and this is distastefully shortsighted and self-centered (or at least human-centered). It fails to convey the vastness of what is."

This is a more subtle assumption than the previous ones. It entails describing the process of cognition (or experience, perception, language) as the projecting of form (boundedness, continuity, causal power) onto what is (intrinsically) independent of our ways of projecting forms. This is somewhat analogous to seeing "what is" as a cloud, and agreeing that though the cloud does not exist as a specific form (as a bird or as a train, even if people can point to it and say, "It looks like a bird!" or "It looks like a train!"), it does still exist as a cloud. This sense is evoked in such phrases as "probability clouds," "space of possibilities," "possible worlds," "collapse of wave functions," or even "oneness," "void," or "being." So, the assumption here is not of the intrinsic existence of either objects or subjects as things, but it is the assumption of some particular kind of something even if absent the qualities that are normally attributed to substance (substantialism). It may equally be the inference of some particular kind of nothing, such as a "void" that exists independent of experiences (nihilism).

Thinking of a meteor hurtling unseen through space, it is easy to think, "It is what it is, no matter what we think or say about it." This is uttered as a testament to the vastness of world, inspiring humility, and safeguarding against an egocentric collapsing of the universe entire onto our narrow short-lived experiences. We go back to thinking about that meteor, so alone, so independent, so capable of being without us, and without any being. "There it is, somewhere out there, unseen." But then here we are, seeing it so clearly, and assuming a great many things about it, despite our very best attempts to see it as maintaining independence from us. "It is what it is no matter what," we say. But we seem to be very clear and convinced about what it is—so much so, that considering an alternative seems perverse. It is definitely a meteor, we might say, a rock and not a god, a collection of atoms and not of thoughts, possessing dimension (even if those boundaries are fuzzy) and hurtling through space (even if space is expressible only in our own experience or language). It is definitely alone and independent—we might go so far as to say it is cold, inanimate, and unfeeling. But how is it capable of being any of these things without us? Where in the meteor itself is the aloneness, the independence, the coldness, nonanimation, unfeelingness? By the same token, where is aloneness, independence, and so forth in the great and vast "what is"? The properties that we assume belong to things in themselves can become as attenuated, subtle, or fuzzy as we can imagine, yet this does not prevent them from being *our* assumptions.

The idea of nonimpingement very quickly and easily becomes accompanied by a host of impingements, such as "what *is* is independent, untroubled, unmoved, unfeeling, cares not for us, is utterly unchanged by our thoughts and lives," and so forth. These are attempts to deanimate the universe, to set it loose so it can be free from the leash of our mere beliefs. However, all of these conceptions powerfully animate and conceptualize the universe as the kind of thing that is capable, by itself, of being independent, unfeeling, uncaring, and so forth.[15]

One can say, "There's something there, it's just that its properties are inexpressible." But even in that thought and statement, the "inexpressible something" is expressed. After all, we are arguing passionately at this point that something is there. Similarly, the statement "There's nothing there," is also very rich with connotation, because we also have very strong ideas about what nothing is. The nothing fills us with a sense of triumph (if we're comfortable with nihilism), or with anxiety (if we're uncomfortable with it).

One can also easily say, "What *is* is completely beyond human conception." But anyone who is truly convinced of this actually stops right there, and does not go on any further to:

1. defend any conception of "*is*-ness" (no matter how subtle, formless, boundless, or "nonsubstancelike"),
2. interpret the "*is*" as having a particular intrinsic property (even the property of being beyond conception),

3. claim that this conception is defensible on the grounds that it corresponds to what *is* (an intrinsically existing "something" or "nothing," however subtle), or

4. take the statement itself too literally by forgetting that this view ("What *is* is completely beyond conception") is itself a conception.[16]

Now, of course, all of the aforementioned statements are contingentist statements. Strangely, therefore, a thoroughly contingentist account (one that takes into account, of course, the contingent nature of the account itself), may actually be quite a thorough and satisfactory statement of non-impingement. If one wants to say something like "It is what it is regardless of our mere beliefs," a contingentist view can help by adding, "If we want to think this way, let's be even more thorough about it by placing this idea of the necessary independence of what is into the category of 'mere belief' as well."[17]

Again, one of the main assumptions commonly made about "what is"—that is, existence itself—is that existence means intrinsic existence: unitary, continuous, and independent. Contingentism offers an alternative to this assumption, suggesting that what it means for the "what is" to exist does not necessitate its radical, necessary independence.

In sum, we tend to assume that the only way of preserving the vastness of what is is by asserting its independence from what is limited, mere, and contingent—that is, from all of these qualities we tend to attribute to our own perspectives, lives, and knowledge. However, that move tends to constrain the very sense of vastness we were trying to preserve, because it makes many assumptions of what it means for there to be "a something there" or "a nothing there." It also makes assumptions about what it means for "what is" to be (that is, the assumption that "what is" has to be independent, that existence means intrinsic existence). Contingentism helps preserve the vastness of what is by questioning these very assumptions. The sense of vastness that contingentism offers should become more vivid in the following chapters, where it should become clear that contingentism implies neither an intrinsically existent human subject that can somehow create a world at will (Chapter 4) nor a conception of organismal knowledge as intrinsically limited (Chapter 5).

What do objects depend on? They depend on observers to bound them and hold them as continuous over time. Their effects depend on observers to distinguish objects from each other, and to note regular interactions between objects that have thus been distinguished. Their properties depend on what is sensed and measured, and on the relations observers make between measurement and theories. Their vividness and their place as bona fide members of the real world depend on the coordination of various bodily movements and sense perceptions, as well as coordination between interacting members of social collectives.

These are several precise senses in which the material world depends on subjects and sensing. These claims neither entail nor produce a world-denying nihilism. They do not lead to some notion of being able to use a kind of "mental substance" (thoughts) to "impact" or rearrange physical substance.[18] And, as we will see in the next three chapters, they also do not in any way lead to the claim that the subject has complete power to custom-make their own isolated world. Rather, the specific claim that the external world depends on subjects and sensing simply calls to attention many processes that give rise to an external world experienced as substances, forces, and things—or even as a subtle, formless "what-is"—where each of these are felt as somehow radically independent of experience.

3

WHAT DOES SENSING DEPEND ON?

Transduction, Energy, and the Meeting of Worlds

As we saw in the last chapter, the notion of the intrinsic existence of objects depends upon a number of assumptions. These assumptions are at play in the standard view of what we call "the sensing of objects" and, relatedly, of what in biology is called "signal transduction." Commonly, it is assumed that objects exist *intrinsically* prior to the sensing activity of organisms, for all the reasons that have been examined before. In addition, at least two more operations occur that make this standard view of sensing seem straightforward:

1. We assume that some things can become the organism (i.e. be *assimilated into* the organism) while others must always remain radically separate from the organism (i.e., be *sensed by* the organism). However, assimilation and sensing can be seen as very similar processes.

2. We assume the intrinsic existence of energy. Even if we see objects and organisms as having no intrinsic existence and as being ever-changing, unfixed forms, we might attribute those changes in forms to something else: the movement of energy. However, energy does not inhere within objects; nor is it a substance; nor is it a force that *makes* things happen in the world through its movements.

In this chapter, I outline the standard view of signal transduction in biology, including a brief intellectual history of the concept of signal transduction itself. I show how the standard model of signal transduction not only *depends* upon the concept of intrinsically existent objects but also actually *reinforces* the very idea that objects are intrinsically existent. I then revisit the concept of signal transduction and offer a contingentist view of the phenomenon that does not rest upon the assumptions of the standard view, and also offers novel and useful reconsiderations of the relation of lower-level and higher-level phenomena; of agency and determinism; of stasis, change, and causal relations; and of the relation of physical and psychological phenomena.

"Signal transduction" is a set of ordered biochemical reactions. It is narrated as a causal story: "This happens, then this, then this." It refers to the process by which an organism converts one type of signal into another. Just as a mechanical transducer, like a light bulb, can convert electrical energy into light energy, similarly a biological signal transducer, like the protein rhodopsin in human eye cells, can convert light energy into chemical energy, and into a signal to which the cell responds.

The simplest signal transduction systems are two-component signal transduction systems (TCSTs), the predominant signal-response systems in bacteria. The "two components" of the system are a sensor protein embedded in the cell membrane and a regulator protein in the cell's interior, or cytoplasm (Figure 3.1). The extracellular portion of the sensor binds or otherwise changes in response to some extracellular signal, inducing the intracellular portion of the sensor to add a phosphoryl group to the regulator protein. The regulator, in turn, changes conformation and stimulates some change in cell state, either by regulating transcription (the production of RNA from DNA, or genes), by modifying enzymes and altering enzymatic activities, or by changing the direction of the flagellar motor.[1]

Though the two components of TCST systems are often tightly coupled, enabling specificity of response to a given stimulus, the components are some-

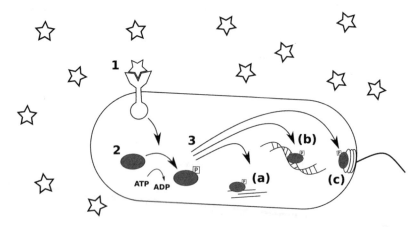

FIGURE 3.1. A canonical two-component signal transduction system. (1) The sensor protein binds a signaling molecule; (2) the sensor adds a phosphoryl group to the regulator protein, which converts ATP to ADP; (3) the regulator initiates a response, such as (a) altering enzymatic activity, (b) regulating transcription, or (c) modulating the turning of the flagellar motor.

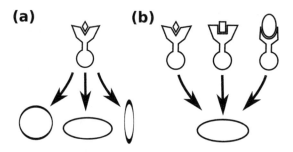

FIGURE 3.2. Convergence and divergence of responses in signal transduction systems. (a) One signal type stimulating one sensor protein type may phosphorylate many regulator protein types. (b) Several signal types stimulating several sensor protein types may phosphorylate one regulator protein type.

times only loosely coupled, enabling cross-talk between various signal-response pathways. As such, TCST systems can enable both specificity and flexibility of responses to stimuli. In the Spo system of *Bacillus subtilis*, which regulates sporulation, a single sensor can phosphorylate multiple response regulators[2]—a case of response *divergence* wherein a single stimulus can elicit multiple responses (Figure 3.2*a*). In the PhoB system in *Escherichia coli*, which regulates phosphate uptake, multiple sensors can phosphorylate a single response regulator[3]—a case of response *convergence* wherein multiple stimuli can elicit the same response (Figure 3.2*b*). Response convergence and divergence represent important deviations from the canonical notion of the TCST system wherein each sensor interacts with a single specific regulator. These "deviations" indicate high potential for novelty in TCST systems—that is, the genesis of new abilities and behaviors using old parts. In the simplest case, a TCST cascade involves four "parts" and three interactions: one stimulus interacting with one sensor interacting with one regulator eliciting one response. However, as noted earlier, at any point in the cascade, the "one" could be "many."

RECEPTORS, TRANSDUCERS, AND SIGNALS: A BRIEF HISTORY

The phrase "signal transduction" first appeared in the biological literature in 1976 in an article that reported that insulin is *produced* but not *released* by rat pancreatic cells when they are in the presence of N-acylglucosamine molecules.[4] It was known at the time that the pancreatic beta cells synthesize insulin in the presence of glucose *and* that these same cells release insulin in the presence of glucose. This led to the question, "How do these cells 'sense' that there's glucose 'out there,' and how do they respond by producing insulin?"

It had been previously hypothesized that both synthesis and secretion of insulin depended on a single glucose receptor.[5] However, Ashcroft and workers found that N-acylglucosamines can stimulate synthesis of insulin, but not its secretion, in the absence of glucose. This indicated that there might be different receptors or different *transducers* for each reaction—one stimulating an "insulin synthesis" response and another stimulating an "insulin secretion" response. They summarized this idea in the following sentence: "It is concluded that there are differences in *signal reception and / or transduction* for the processes of insulin biosynthesis and release (emphasis mine)." These phrases—"signal reception and / or transduction"—were uncommon in the 1970s. However, the language of "receptors" was widely used, in such terms as "glucose receptor," meaning a protein that binds glucose. When did the "receptor" term arise and become common? How did biologists come to speak of cells having external receptors for various substances?

In the late nineteenth century, pharmacology was emerging as a scientific discipline. Though drugs had been used in various medical practices for millennia, mechanistic accounts about the specific reactions of drugs in the body became more highly valued. At the time, the ascendant scientific view was that drugs acted in the body by stimulating nerve cells through their nerve endings.[6] Physiologist John Newport Langley's research undermined this view by demonstrating that nerve cells without nerve endings could still be stimulated. In 1905, he originated the concept of a "receptive substance" in cells that reacts with external chemicals.[7] Meanwhile, Paul Ehrlich, considered "the father of immunology," studied the selective uptake of dye by different cells, and hypothesized that "side chain molecules" on cells react differentially with dyes (Figure 3.3). He replaced the term "side chain" with the term "receptor" in 1900.[8] Like a number of biologists of their time, both Langley and Ehrlich conceived of the cell as being a very large, complex molecule called "the protoplasmic molecule," a term coined by physiologist Eduard Pfluger in the 1870s, and later called *biogen* by Max Verworn.[9] The protoplasmic molecule had a core structure that remained stable, and a number of side chains that reacted with various chemicals. Thus, the stability of cells was explained by the backbone, and their dynamism or reactivity was explained by the side chains. "Animal oxidation," as it was called, somewhat analogously to what is now called "cellular respiration," was explained by the reactivity of side chains with oxygen, thus explaining why animal tissues consume oxygen.[10] Ehrlich's expansion of the types and functions of side chains—some reacting with oxygen, others with alkaloid dyes, and yet others with drugs and toxins—was thus continuous with other types of cellular reactions within this paradigm.

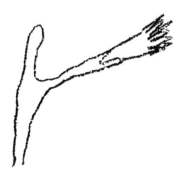

FIGURE 3.3. Paul Ehrlich's model of an amboceptor, a type of side chain of the protoplasmic molecule capable of binding substances. Here a substance is shown bound to one of the side chain's two binding sites. Sketch by Paul Ehrlich, 1901. Courtesy of the Rockefeller Archive Center, New York.

The receptor theory was not particularly popular at first. Physiologists like Walther Straub believed that chemicals like toxins and drugs affected cells not by reacting with cells, but by physically surrounding cells and blocking their ability to excrete wastes, thereby damaging or simply altering cellular activity.[11] In the 1930s, pharmacologist A. J. Clark sought to describe what laws of "physical chemistry" might be "postulated regarding the combination formed between drugs and cells."[12] He calculated that drugs are effective at a concentration lower than what would be required to surround cells. He also found the mathematical relationship between drug concentration and cell response to be a hyperbolic curve, like the relationship of the adsorption of gas molecules to a metal surface. He concluded that the mode of drug action was a kind of adsorption reaction where molecules bound reversibly to single receptors.

Langley, Ehrlich, and Clark were pioneers of receptor theory. However, the theory was not more widely accepted until much later, after Raymond Ahlquist's 1948 study of the inhibitory versus excitatory effects of adrenalin on different cell types. Ahlquist attributed the two different kinds of effects to the relative distributions of two different adrenalin receptor types, one of which could be blocked by an adrenalin antagonist, thereby inhibiting one response.[13] This brings us back to the 1976 article by Ashcroft, Crossley, and Crossley, who thirty years after Ahlquist, used the word "transduce" and the word "signal." They suggested that *receptors*—or whatever "receives" a chemical—are distinct from *transducers*, or what "converts" that chemical into another form (a "response").

Though earlier work on receptors took place in animal cell systems, advances in microbiology were important in developing a signal transduction concept. In 1977, around the same time as Ashcroft and co-workers published their findings on insulin in mammalian cells, neurophysiologist Eilo Hildebrand wrote about signal transduction versus energy transduction in the microorganism *Halobacterium*. It had been known that the bacteriorhodopsin protein captures light just like the rhodopsin protein in human eye cells, and it had been suggested that *Halobacterium* bacteriorhodopsin be used as a model system in order to gain insight into the process of human vision, and particularly the role of rhodopsins ("light-capturing molecules") in vision. Though it had been known that unlike in human cells, bacteriorhodopsin converts light energy to chemical energy in the form of ATP,[14] Hildebrand asked whether bacteriorhodopsin *also* played a role as a "photosensory mechanism," mediating *Halobacterium* cells' movement away from light. This relationship between signal and energy transduction is an important insight that will be discussed further below.

The phrase "signal transduction" gained greatly in popularity in the 1980s (as illustrated in Figure 3.4), after Martin Rodbell's development of the G-protein coupled receptor (GPCR) concept in the mid-1980s.[15] GPCRs are the most numerous and most important sensory signal transduction systems in eukaryotes. They are found in a great many mammalian cell types and are important elements in the signal transduction events necessary for all five human senses. Rodbell described his model for the transducer as highly influenced by Norbert Wiener's cybernetic theory—that is, a theory of feedback, and of how systems, living and nonliving, use feedback.[16] Rodbell's transducer was conceived of in abstract cybernetic terms: discriminator, transducer, and amplifier.[17] In a cybernetic framework, the cell is conceived much like a microphone, or like a thermostat, or like a water clock—mechanical and self-regulating. The discriminator receives specific types of signals, the amplifier turns the signal into a larger-scale response that affects the entire cell, and the transducer is "a coupling device designed to allow communication between discriminator and amplifier."[18]

Even the brief and sparse history narrated here regarding the development of the signal transduction concept from the late nineteenth century to the present reveals in part its contingent existence. Through this history, three basic notions of what the cell is, and how its responsiveness[19] to external conditions arises, can be discerned.

1. Ehrlich, Langley, and many contemporaries (late nineteenth century to early 1920s) describe the cell as a complex chemical ("protoplasmic molecule") that has a stable core structure and reactive side chains. Cell-environment interactions are described as reactions of these side chains with nonliving chemicals. There is no mention of "sensing" per se, just "reactivity" and "reactive

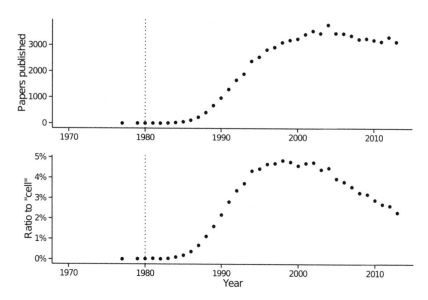

FIGURE 3.4. Number of papers published in a given year (a) with the term "signal transduction" appearing in the title or abstract and (b) with the term "signal transduction" normalized to the number of papers with the term "cell" in the title or abstract (to adjust for increases in the number of total papers published over time). To normalize, the number of papers with "signal transduction" was divided by the number of papers with "cell" in title or abstract for each year. Data source: PubMed.

substance." Metabolism and pharmacological reactions are the same in principle. There is no strong divide between chemical and biological reactions. As Hammarsten and Mandel wrote in 1904, "The living protoplasmic molecule differs from the ordinary non living proteid ('complex chemicals found outside the organism or in the animal fluids') by being more unstable and therefore having a greater inclination toward intramolecular changes of the atoms."[20] Thus, interestingly, the main difference between biological and chemical substances is primarily the former's "instability." Ehrlich's illustration (Figure 3.3) suggests the protoplasmic molecule as a single structure with many branches or protrusions.

2. Hildebrand and other microbiologists studying free-living single cells described cells as organisms—something like "tiny animals," or Antonie van Leeuwenhoek's "wee animacules." Cell-environment interactions are described as "sensory" and cells are described as having "behaviors," and even making "decisions." In 1973, Julius Adler, a microbiologist and a major pioneer of research in the movement of bacteria in response to various chemicals, published a paper entitled "'Decision making' in bacteria." In this paper, Adler set

up an experimental "'conflict' situation [where] a bacterium must 'decide' whether to pursue the attractant, ordinarily a nutritious chemical, or flee from the repellent, ordinarily a harmful chemical."[21]

3. Rodbell and other cell physiologists described cells as cybernetic systems, mechanical and self-regulating like a thermostat. Cell-environment interactions are described as feedback loops, where external perturbations trigger reactions that either amplify or dampen the perturbation. In his 1948 classic *Cybernetics*, Norbert Wiener, considered the father of the field of cybernetics, wrote, "Information is information, not matter or energy."[22] Information is thus described as its own quantity, and sensing is a kind of "information processing." These concepts have echoes in contemporary systems biology and in cognitive science.

Thus, depending on one's conceptual framework and explanatory goals, the responsiveness (or reactivity, activity, irritability, dynamism, behavior, and so forth) of cells can be thought of as an instantiation of some more general principle. For example, as suggested in the preceding list, cellular responsiveness is a kind of chemical reaction, or a kind of behavioral response, or a kind of information processing. Just by considering these three alternatives, one might glimpse the vastness of the descriptive and explanatory horizon available for describing, explaining, and understanding cells and cellular activity. Cellular responsiveness might be described as a kind of "just-about-anything"—a kind of fluid wave, a kind of hydraulic system, a kind of change in frequency, a kind of creative expression, or a kind of physical decay. Each of these conceptions would be coherent and justified within some paradigm or, as immunologist Ludwik Fleck wrote, "thought style."[23] As commonsensical as the most popular contemporary notions of cells may seem, no conception of a cell's statics or dynamics captures or exhausts what it *is*. Once again, the question of "is" or existence arises. I will return to it after considering an interesting meeting of the concepts of "transduction" and "sensing" in the 1970s.

SIGNAL TRANSDUCTION AND CELL SENSING

Eilo Hildebrand, a neurophysiologist studying mammalian vision, looked to a unicellular organism to gain insight into phototransduction, or the conversion of light energy into another form. The unicellular system he studied was the archaeal microorganism *Halobacterium salinarum*. In a 1977 paper entitled "What does *Halobacterium* tell us about photoreception?" Hildebrand expressed a sense of relationship between signal transduction and energy transduction in describing two putative functions of bacteriorhodopsin in *Halobacterium* cells:

Since the discovery of bacteriorhodopsin (BR) by D. Oesterhelt and W. Stoeckenius (1971) it has been expected that *Halobacterium* could serve as a photoreceptor model which would help to understand the process of vision. Several experiments brought strong evidence that BR functions as an *energy transducer* converting light by means of a light-driven proton pump into chemical energy which is finally stored in the form of ATP (Oesterhelt and Stoeckenius, 1973; Oesterhelt and Hess, 1973; Danon and Stoeckenius, 1974). Thus the main function of BR is quite different from that of rhodopsin in visual photoreceptors which act as *signal transducers* transmitting information from the environment to the nervous system.

However, the bacteria also show light-induced behavioral responses and one could presume that BR, besides its energy converting function, might be involved in a *photosensory* mechanism. First experiments done by H. Berg and K. Foster using a three-dimensional tracking microscope indicated the existence of two photosystems in *Halobacterium* which trigger motor responses. One of them seemed indeed to be related to BR (Berg, personal communication, 1973). Thus the question became again interesting whether *Halobacterium* could be regarded as a model system suitable to study the *transduction mechanism of photoreceptors*, and N. Dencher in our laboratory started to analyse the photoresponses in detail. (emphases mine)

There are a number of crucial insights in Hildebrand's account. First, it illustrates the ambiguities that arise when using unicellular organisms as model systems, particularly to gain insight into something like the sensory, cognitive, or behavioral processes of multicellular organisms. There is a basic ambiguity of working with unicellular organisms and generalizing from our insights about them, for we do not always know whether to see them primarily as *cells* and therefore analogous to cells within a multicellular organism, or as *organisms* and therefore analogous to multicellular organisms themselves. This is not necessarily a problem of miscategorization, for unicellular organisms are, of course, both cells *and* organisms. However, it can lead to confusion when deciding just what a unicellular model organism is a model *for* in any particular case.

The concept of the unicellular organism as a *model system* is itself a historical development in biology. The idea that unicellular organisms provide insights into the cell processes of multicellular organisms and into the lives of multicellular organisms themselves is now a very common and useful one in biology, but it is nontrivial. The model organism approach depends on a view of sameness between organisms and the ability to generalize across life forms. Charles Darwin's theory of heritability and descent from a common ancestor provides much of the basis for this sense of commonality, certainly, but so does the ascendancy of research questions that can be answered on a molecular or biochemical level. If we are looking at the level of biochemistry, we might say that organisms really are very much the same. It is a little like taking a sand grain as a "model organism" to gain insights into mountains or houses—at the level of chemistry, they are very much the same as well. Microorganismal model systems have been incredibly useful in exploring some important and widespread biological processes. The light and dark reactions of photosynthesis were analyzed using the motile unicellular algae *Chlamydomonas reinhardtii* (illustrated in Figure 3.1), the eukaryotic cell cycle and hormone responsiveness were analyzed using the brewer's yeast *Saccharomyces cerevisiae*, and major processes of metabolism and of genetic regulation were analyzed using the bacterium *Escherichia coli*, to name just a few famous processes and microorganisms. The successes of researchers in working out all of these processes in unicellular organisms is part of what motivated Hildebrand and others to look to the archaeal species *Halobacterium salinarum* to gain insight into photoreceptors in mammalian (notably human) systems. Where it gets interesting is in considering whether there could be such a thing as a unicellular model system to explore sensory processes, behavior, or even cognition. The answer depends on (1) how one defines these vexed terms, (2) the extent to which one believes these processes to be primarily small-scale and local, that is, traceable to the single cell, (3) the extent to which one believes these processes to be explicable in the language of biochemistry, and (4) the extent to which one takes the single cell, as a "basic unit of life," to have all of the most important "properties" of life of all kinds and all scales, including responsiveness, behavior, and perhaps cognition.

Hildebrand looked for the link between light and "behavioral responses" in cells. A motile cell's erratic and unpredictable movements often contradict expectations for the more simple-seeming movements of machines, which may be why the word "behavior" is used instead of "motion."[24] "Behavior" also suggests a functional aspect to movement—"the cells move," we might say, "*in order to* avoid or get closer to light." Interestingly, it is the noticing of the behavioral responses (movement) of cells away from light that made Hildebrand and colleagues again take up the question of whether *Halobacterium* cells would provide some insight into light transduction in mammalian photoreceptors. However, mammalian photoreceptors do not themselves move in response to light—it was known that they undergo a membrane depolarization (i.e., the cells become less negatively charged) but remain in place. Signal transduction could easily be taking place in *Halobacterium* without cell movement. This suggests that Hildebrand was looking to the cells as analogous to a multicellular organism—a whole organism that *behaves* a certain way in the presence of an external stimulus.

Looking for this link between light and behavior, Hildebrand goes on to write, "Because of its small size it seems hopeless to insert a microelectrode for intracellular voltage recording, so that the application of this classical electrophysiological method is out of the question." He is referring to the methods used to measure membrane depolarization[25] in mammalian nerve cells, including retinal cells, indicating that the *Halobacterium* cell is considered analogous to the retinal cell. However, in stating that the *Halobacterium* cell has a "photosensory" response, Hildebrand is subtly making an analogy to a whole multicellular organism. The single cell transduces light and moves in the presence of light. A mammalian photoreceptor cell does not move in response to light, and any movement that may occur as a behavioral response to light is an *organismal* response, dependent on the single cell being integrated into a larger system. It seems Hildebrand wanted to see whether the light induced a depolarization in *Halobacterium* cells, as it does in mammalian photoreceptor cells. Furthermore, he wanted to find whether that depolarization was related to the cell's movement. Today, there is no indication that such electrical signals occur on the level of a single *Halobacterium* cell. Rather, *Halobacterium* phototaxis occurs much as *E. coli* chemotaxis does—through a cascade of protein interactions that increase the probability that more cells out of a given population will move away from light rather than toward it (Figure 3.5). What links stimulus and behavior is thus not an "electricity"-based "mini nervous system" in a cell, but conformational changes in molecules which induce conformational changes in other molecules. This direct interaction of molecules is made possible by diffusion, which occurs quickly on microscopic scales, but very slowly on larger scales.

Hildebrand uses the concept of the transducer as a structural component of a cell, but *signal* transducer or *energy* transducer as a functional designation. In other words, some protein (in this case, bacteriorhodopsin) responds to light and can function (1) as an energy transducer, by "converting light by means of a light-driven proton pump into chemical energy which is finally stored in the form of ATP" or (2) as a signal transducer by converting light into some signal, e.g. into electricity, or into a change in conformation of some molecule. In the case of mammalian rhodopsin found in cells in the retina, we might say that rhodopsin converts light into electricity, which is integrated with other electrical impulses in the brain and is involved in vision.

Hildebrand's use of the word "photosensory" is interesting in this regard. "Phototaxis" is a more conservative term, simply meaning "movement in the presence of light." "Photosensation" may have some additional implication of

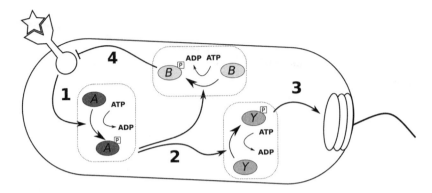

FIGURE 3.5. The chemotaxis pathway of *Esherichia coli* and *Salmonella typhinurium*. (1) A transmembrane receptor binds to a chemoattractant (star) and activates the kinase CheA, which in turn (b) phosphorylates the protein CheY, which (c) controls the direction of the rotation of the motor proteins that drive the flagellum. CheA also (d) phosphlorylates the protein CheB. Phosphorylated CheB modifies the transmembrane receptor to make it less sensitive to the chemoattractant, which allows the cell to adapt to background levels of chemoattractant.

cognition. If we see a human being walking toward light, based on our own experiences, we often infer that there is some awareness of the light and some motivated movement in that direction. What is happening in the cellular case? Is there a sensing awareness or a motivated agent there? In other words, does a cell "see" light, or does it simply move in the presence of light? What can studying the cell tell us, if anything, about how *we* see light?

ASSUMPTION OF SAMENESS AND DIFFERENCE

We can look to plants for some insight into this question. A plant undergoing photosynthesis is said to *assimilate* light, analogous to the way a primate might assimilate an apple. That is, the plant depends on light to create sugars, which in turn become the body of the plant. We generally think of chlorophyll as involved in plants *using* light in this way, but not in *sensing* light, and certainly not in *seeing* light. However, chlorophyll derivatives have also been found in animal eyes, where they *are* involved in the sensing and seeing of light. In the photoreceptor cells of certain fish,[26] chlorophyll absorbs and transfers photons to the primary visual pigment (retinal) (Figure 3.6), which changes conformation and initiates a cascade of molecular conformational changes. In this context, chlorophyll is called a "photosensitizer" because it captures light of a certain wavelength that would not normally be captured by the primary visual

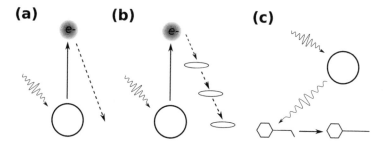

(a)   **(b)**   **(c)**

FIGURE 3.6. Chlorophyll in various contexts. (a) Electrons (e-) in the chlorophyll molecule (circle) absorb photons (wavy arrow) and jump to a higher energy level (solid arrow). That energy can be rapidly dissipated as heat or light (dotted arrow). When a solution of free chlorophyll is exposed to light, the solution warms and becomes faintly fluorescent. (b) Chlorophyll in plants is coupled to an electron acceptor (oval), which is coupled to an even stronger electron acceptor, and so forth. When electrons in the chlorophyll are excited by light, the electrons are passed from one electron acceptor to the next. This prevents the rapid dissipation of energy in the form of heat and light. (c) Chlorophyll derivatives are also found as *visual* pigments (photosensitizers) in the retinal cells of some fish. Again, electrons in the chlorophyll molecule absorb photons (short wavy arrow) and are excited, but here they emit new, lower-energy photons (longer wavy arrow), which are transferred to *cis*-retinal molecules. When photons hit *cis*-retinal (hexagon with kinked tail), the molecule changes conformation to its all-*trans* form (hexagon with straight tail). This sets off a series of conformational changes in various molecules downstream involved in vision.

pigment and transfers it to that pigment, thus expanding the total spectrum of light that can be sensed or "seen." In fact, studies are underway to see if chlorophyll can be used as a photosensitizer in mammalian systems, extending vision in the far-red spectrum. Studies of mice injected with a chlorophyll derivative show some evidence of increased responsiveness to red light.[27]

If a chlorophyll solution is kept in a vial, the chlorophyll is excited by photons, emits electrons, and this energy is dissipated as heat and light. The solution glows faintly and warms up (Figure 3.6a). In a chloroplast, instead of the energy being dissipated as light, it is harnessed via an electron transfer chain, and ATP is produced. This is classic photosynthesis (Figure 3.6b). In dragonfish eyes, the chlorophyll emits a lower-energy photon, which is transferred to a *cis*-retinal molecule, changing it to an all-*trans* form, and this conformational change initiates, in some sense,[28] a whole visual signal transduction cascade (Figure 3.6c). (Interestingly, dragonfish do not synthesize chlorophyll like plants do; they did not evolve "chlorophyll genes" that were selected for. Rather, chlorophyll that dragonfish get through their diet is incorporated into the eyes.[29])

These look like very similar processes. Yet, sensing and metabolism are often conceived of as very different. Why is this? One explanation is that this sense of difference arises from the very notion of a "transducer."

A transducer is a rich idea. In engineering, it is something that converts one form of energy into another. Radios transduce radio waves into sound energy, and light bulbs transduce electrical energy into light energy. When energy (or signals, or matter) enters a transducer, it comes out differently on the other side. In this sense, the transducer is a kind of convenient black box—something happens on one side of it, and something happens on the other side of it, but if we do not want to, we do not have to look inside of it to see what happened "inside" of it. For many explanations, it suffices to black box the transducer and simply say, "it converts one form of energy into another."

The trick with the concept of signal transduction and of an object called a transducer is that *it stabilizes the sense that whatever stays outside is different from what comes inside.* To express this in more "process" terms, whatever happens outside is not the same as what happens inside. A sugar that does not cross a cell membrane but binds a receptor (or sensor) on one side of the membrane *is radically different from* the changes in the receptor's conformation following the binding, the chemical cascade initiated by the conformational change, and the cell locomotion that results from the cascade. We might ask, "What makes the magic happen? What is in the 'black box' of the transducer that keeps the external world on the outside, yet produces some *effect* on the inside?"

There are many kinds of ways in which molecules come into cells: through ion channels (Figure 3.7a), by binding to transporter proteins (Figure 3.7b), by being incorporated into vesicles (Figure 3.7c), and by diffusing through the membrane (Figure 3.7d). In all of these processes, there is the sense that whatever was outside of the cell *is the same* as whatever comes into the cell. Interestingly, this may be true even in cases where a substance is modified in its entry into the cell, for example, in the case of certain glucose transporters that add a phosphoryl group to a glucose molecule as it enters the cell, thus creating glucose–6-phosphate. Glucose was a thing on the outside of the cell, and what enters the cell is *the same thing, slightly changed.* Language is very helpful in confirming the sense of sameness in this case: "glucose–6-phosphate" is a wonderful way of linguistically simply attaching something (phosphate) onto the same thing (glucose). In the case of certain sucrose transporters that cleave sucrose into glucose and fructose upon transport, it may be more difficult to conceive of what was outside and what comes inside as the same thing, but we still call it "the same thing, broken down."

What makes it possible to say that molecules are the same even as they are modified through biological processes? It is a kind of subtle essentialism—the

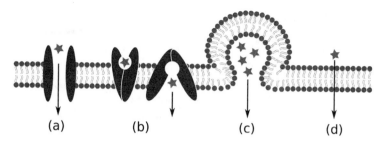

FIGURE 3.7. Movement of molecules into cells. (a) Ion channels, (b) transporter proteins, (c) vesicles, (d) diffusion.

implicit claim that an object can be the same thing even though it changes form, as if there was some essence within the object that makes it what it is throughout those changes.

To be clear, there is no practical problem with using conventional biochemical language. "Glucose turning into glucose–6-phosphate" and the rest of it is a way of speaking that is very useful for many purposes. However, concealed in this formulation is some implicit assumption about what counts as part of the organism, and what counts as external. When the world on the outside is considered to directly enter the world on the inside, this process is commonly called "assimilation." Light that is assimilated into a plant in the case of photosynthesis changes the physiology of the organism and actually *constitutes* or *builds* the organism. In the case of sensing, it can seem as though the world—particularly as *objects*—stays on the outside of organisms. Why do we not see sensing as a process of assimilation as well? Sensing is, after all, also a process that constitutes and builds organisms.

Take, for example, the question of a curious child: "When does food stop being food and become *me*?" One might say to such a child, "Well, sweetheart, we don't exactly know." Holding an apple in one's hand, the apple is obviously not the hand. Having chewed, swallowed, and digested the apple, one cannot be so sure that the apple is not the hand. The apple can be made the same as the body.

However, it seems that most things cannot be made the same as the body. Looking at an apple is not the same as eating an apple. Touching a pen and writing with a pen—this indigestible object, this thing that one could not make a part of one's body even if one wanted to—is not assimilating a pen. Watching the road is not the same as making the road a part of one's body. Or is it?

We already know from our everyday lives that sensed objects are assimilated into the body. The smell of food produces saliva as surely as the ingestion of food produces gastric juices. Scary movies elevate heart rate, as does the ingestion of caffeine. Even the most transitory sense experiences can change the body on much longer time scale. A sudden loud sound can be incorporated into the body as muscle tension that lasts for minutes, hours, or days. Moreover, the sound can produce muscle tension and then a change in posture that becomes habitual, such that the muscle tension lasts on the order of years. Such effects are just as significant as metabolic assimilations, such as when glucose is incorporated into the body as glycogen in the liver.

Sensory events are typically described as the body's *responses* to external stimuli. After all, it is not as if by looking at an apple, the apple dissolves into a cavity under my skin. However, in a nontrivial sense, the apple really does become a part of my body; upon seeing it, my physiology is not the same. Moreover, this act of seeing itself is not strictly a response to what is outside of me. I am responding to the experience that is arising dependent upon my cognitive processes (as discussed earlier: bounding, imputing continuity, experiencing coordination of senses, etc.). I am also responding to my own *prior* experiences, where this experience of "apple" is not at all experienced as unique and unrepeated but is related to past experiences of apples (my stomach gives an involuntary turn and my throat constricts without warning, as my body re-experiences an apple I ate once—along with the worm in it). All of this arises from seeing what is ostensibly radically outside of me.

The intuition that "information" exists intrinsically and is *transferred* between things without itself being changed is part of what both creates and perpetuates the idea of sensing as "*not* assimilation." For example, when is light considered "energy to be used," and when is it considered *information* about the state of the external world? The chlorophyll example illustrates that there is no inherent distinction between "the use of light energy to perform work" and "the use of light as information" or "the use of light to guide behavior." Both signaling and metabolism are conversions of energy (though, as we will see, energy itself is not inherently existent).

"Information" terminology can be fraught with difficulties. Information has something to do with the abstract taking form—hence the Latin *in* or "into" plus *forma* or "form." In *The Ontogeny of Information*, Susan Oyama challenges the idea that the information which creates the living organism somehow exists *before* the living organism itself exists. She systematically analyzes and rejects the notion that there is "genetic information" that somehow "encodes" the living organism, a kind of abstract form that precedes physical form. Just as we sometimes conceive of DNA as "genetic information" that is transmitted from generation to generation and "translated" into living bodies, similarly, we conceive of "external information" or "information *about* the world" as transmitted to the organism and translated into responses, behavior, or cognition. In both cases, the idea is that information is somehow "out there," a form before form, which exists prior to the actual bodies and behaviors of organisms—a nonempirical, metaphysical view.

In the next chapter, I examine the question "What is external to the organism?" by asking, "What are the spatial boundaries of the organism? Do organisms end with the membrane, the skin, or the like?" However, the present discussion of transduction presents another angle from which to address the question of what is external to organisms. It is not simply about the delineation of a spatial inside-outside boundary, or about the recognition that such a boundary is not fixed or inherently existent. The question of what is external to organisms is also dependent upon notions of sameness and difference over time.

To return to the original motivating question of this discussion, do plants (merely) use light for energy, or do they sense light? These processes are only held in contrast with one another if one assumes that in the first case, the light is made the same as the organism, and in the second case, the light is kept separate from (outside of) the organism yet produces a change in the organism. That such divisions are commonly made reflects a long history in biology and cognitive science of taking bodily processes to be not only radically different from but also *lower than* cognitive processes.[30] However, when such divisions are not made, and when sensing is not assumed to be a process that keeps objects on the outside of organisms, new views of sensing can arise, as we will later discuss.

ASSUMPTION OF ENERGY AS A KIND OF SUBSTANCE

Another concept that is central to the idea of the transducer and the concept of signal transduction is that of *energy*. After all, transduction is said to be the conversion of energy from one form to another. Moreover, not only can energy be said to "be converted," and to "flow" between objects, these objects are themselves said to contain energy. For example, as physicist David Jeffery writes in his lecture notes for an introductory physics class, "energy is associated with the chemical bonds of the atoms in gasoline. So this chemical energy is associated with the relationship of the atoms. But most often you like to say that there is energy in your gallon of gasoline."[31]

What is energy? Is it in objects, and does it flow through objects? Physicist Richard Feynman describes energy as follows:

There is a fact, or if you wish, a law, governing natural phenomena that are known to date. There is no known exception to this law; it is exact, so far we know. The law is called conservation of energy; it states that there is a certain quantity, which we call energy, that does not change in manifold changes which nature undergoes. That is a most abstract idea, because it is a mathematical principle; it says that there is a numerical quantity, which does not change when something happens. It is not a description of a mechanism, or anything concrete; it is just a strange fact that we can calculate some number, and when we finish watching nature go through her tricks and calculate the number again, it is the same.[32]

It is easy to forget just how *abstract* a concept energy is. Everywhere and always, forms are changing, and we want to be able to attribute those changes to some *thing*—something concrete, forceful, and substantial. The word "energy" is often used in that way. For example, to revisit our earlier example, when light strikes a retinal molecule in your eyes, the retinal molecule changes shape (see Figure 3.6c). We say that energy was transferred from light to the retinal, inducing a conformational change. If we analyze more closely, we can see the cognitive work that makes such a statement seem straightforward.

First, the retinal is "held still" as an object by an observer. It is called the same object before and after the change, and it is said to isomerize, or "change form." This is a subtle essentialism—the idea that an object can change form and still be the same object, as if there was some essence to the object that makes "the object" different from "its form." This is analogous to thinking that there is some essence in a flower that makes the blooming flower and the withering flower the same flower.

Second, those changes in the object's form are attributed to the absorption of energy. According to quantum electrodynamic theory, an electron absorbs a photon and spins at a higher orbital, which makes it "unstable" in that it becomes more attracted to the protons of adjacent atoms than to the protons of "its own" atom. When electrons switch atoms, the dynamically sustained pattern of attraction and repulsion that made the molecule seem stable is disrupted, and a conformational change in the molecule occurs. However, though it is often assumed that energy is what makes things happen in the world, and energy is understood at least metaphorically as a forceful substance, the metaphor breaks down. Energy is neither a powerful object nor a fluid substance. Energy is a description of a set of precise regularities, not an agent, a thing, a mechanism, or a causal power. The energy that we might want to evoke as a power or a causal agent that made something happen in the molecule is a *number*. If we define "the system" carefully as a closed system including the light source itself, the eye, and the photoreceptor cells, energy is a number that will have stayed the same before and after all the reactions that took place from the emission of the photon from the light source to the conformational change of the retinal.

Third, we say that there are changes in the *forms* of energy—from light energy to chemical energy, in this case—but continuity of energy *itself.* This is analogous to the idea that an object can change form, but remain the same object: It is assumed that energy can change form but remain the same energy. If energy is considered a *substance*, then this is a kind of essentialism. However, if energy is considered a *number*, there is no essentialism. The number is contingently existent, dependent on an observer to bound the system a certain way.

What we see, then, is a very subtle relationship between change and stasis. One says that the object stays the *same* but *changes* form. The *change* in form is attributed to a *change* in forms of energy, but energy as a quantity itself stays the *same*. Importantly, all attributions of sameness or difference over time are observer-dependent. The observer who watches the changes or does the math determines what will be held still as an object or substance and what will be seen to change in contrast to that stillness.

This brings us back to the contingentist's refusal to accept the inherent sameness or difference of things. If we see the world as things, objects will be "the same substance over time undergoing changes in form," and energy will be "what causes changes in object-forms." If we see the world as energy, energy will be "the same substance over time undergoing changes in form" (for example, energy changing from its light-energy form to its chemical-energy form), and objects will be "what cause changes in energy-form" (for example, retinal molecules or solar panels which change light energy to chemical energy) (Figure 3.6c). In either case, the confusion is created by thinking of energy and form as *separable* from one another, when they are not. The relationship between energy and form can therefore be understood coherently only in contingentist terms. Even then, "understood" may be too strong a term. The contingentist views the relationship of energy and form in a careful way that does not separate them into two "things," and thus gestures at the subtlety of the relationship.

Again, the suggestion here is *not* that energy is not a very useful concept, nor that it does not help illustrate precise regularities. The suggestion is that in conceiving of energy as being transferred between things, one can miss the subtlety of the phenomenon. It is a little like thinking that because you can read these words, the words are actually being *transferred* from this page to your eyes right now. However, in either case, no thing, no matter how subtle, is being transferred.

In sum, things are said to change from one form to another, and this change in form is attributed to the movement of energy. Energy, however, is not a forceful substance that makes things happen to objects, but it is a (verbal or mathematical) term that is often evoked precisely to explain and predict changes in forms. When we do not evoke "transfers" of energy or information, we are left with new possibilities of how to account for the changing of phenomena, as will be discussed at the end of this chapter.

RELATING PHYSICAL AND PSYCHOLOGICAL PHENOMENA

We may return now to Eilo Hildebrand's problem of trying to explain the relation between the *reactions* of chemicals and the *responses* of microorganisms. In the standard signal transduction account, "sensing and responding" means "undergoing an ordered set of chemical reactions"—a signal transduction event. On the other hand, "sensing and responding" can also mean "perceiving and acting"—something that whole organisms (and not mere chemicals) are said to do. "Sensing and responding" is thus a metaphor with two senses. In one use of the metaphor, sensing and responding is *like* materials undergoing transformation in a chemical reaction (like iron reacting with oxygen and transforming to rust, a physical-chemical phenomenon). In another use of the metaphor, sensing and responding is *like* unified agents with choice perceiving and acting (like our own experience of perceiving an intersection and deciding to go left or right, a psychological-behavioral phenomenon). In short, the metaphors we have for describing signal transduction are the *physical* language of objects causing transformations in other objects (chemical *reactions*), and the *psychological* language of agents sensing, deciding, and behaving (biological *responses*). The ambiguity arising from these two uses of the metaphor can be expressed in at least three alternative ways:

1. Both cells and animals are responsive.
    (a) Cells are like animals that perceive and act.
    (b) Animals perceive and act dependent upon cells that perceive and act.
2. Cells are reactive, and animals are responsive.
    (a) Cells are *not* like animals that perceive and act. Cells (merely) undergo chemical reactions.
    (b) Animals perceive and act dependent upon cells that undergo chemical reactions.
3. Both cells and animals are reactive.
    (a) Cells are like animals that undergo chemical reactions.
    (b) Animals undergo chemical reactions dependent upon cells that undergo chemical reactions.

Part of the ambiguity here arises due to assumptions that are commonly made when considering the relation of smaller-scale and larger-scale phenomena. This relation is a type of causal relation. One phenomenon is said to cause another, but in the contingentist view, no intrinsically existent causal power links one phenomenon to another. Rather, in the absence of certain conditions, other conditions do not arise—yet there is no *link* between the two other than the cognitive process of linking. A similar cognitive process occurs in linking smaller-scale and larger-scale phenomena. A whole is caused by, or depends upon, its parts. Without the parts, there is no whole. Wooden chairs depend upon wood itself, copper wires depend upon copper atoms, cells depend upon molecules, and animals depend on cells. Yet, in the contingentist view, there is no *link* between the smaller-level and the larger-level phenomena, except for the cognitive process of linking the two. Whenever we observe a whole, we can always dissect it and discern an often predictable set of parts. Still, the assemblage of parts doesn't somehow *make* the whole happen. That is, the parts have no inherent *power* to bring about the whole. This point was also discussed in a different context in the previous chapter, where the concept of "emergent properties" was analyzed. The whole has different properties from the parts, yes, but not because some novel property intrinsic to the new whole suddenly *emerges* from the assemblage of parts. Rather, it is because a whole is related with *as* a whole that it is observed to have particular properties that are not present when it is related with as a set of parts.

One can say, "Copper is copper because it's made of copper atoms, and not, say, aluminum atoms, helium atoms, or bits of cheese." However, it is precisely what has the (observable) properties of copper that is said to be made of copper atoms. If we saw that the thing appeared to every observation as a block of aluminum, we would say, either by positing and refining over time a theory of aluminum particles, or by empirically dividing the aluminum into small particles, that the thing was made of aluminum particles. In the case of one, we always find the other, but one does not *make* the other happen. For those who view phenomena from a contingentist perspective, the idea that what is smaller is also more *basic* simply does not arise. What is smaller is only smaller, and arises only from the cognitive relating of "a whole" and "parts."

Are we (merely) using two different languages to describe the same thing, or, as we like to say in the spirit of pluralism, are we seeing the same thing in two different ways? Not exactly, for larger and smaller scales are *not the same thing*. A cell is a cell. A set of molecules is a set of molecules. There is the sense that the two are interrelated, of course, but that interrelation is, as in any causal account, the cognitive linking of one with the other. A set of molecules does not precisely "become" a cell, nor does a cell "emerge" at a higher level of organization *from* a set of molecules. Rather, the cell is a cell when related with as a cell, and a set of molecules is a set of molecules when related with as such.

Therefore, contingentism is not commensurate with a reductionist view of biological phenomena simply because it does not assume the intrinsic existence of a quantity or a phenomenon to which everything can be reduced.[33] The world is not intrinsically *one* quantity—such as particles, forces, or energies—that are then experienced by organisms as *many* quantities such as sugars and salts, shoes and stones. A multifarious world is *genuinely* multifarious precisely because it is experienced as such. There is no intrinsically existent oneness to which this multiplicity is reducible. No organism relates with a reduced world. Reductionism is a sophisticated intellectual *operation* that need not lead to claims about an intrinsically existent state of affairs. A world is not intrinsically particles, energy, and forces any more than it is intrinsically earth, fire, and water, or animals, minerals, and vegetables. On the other hand, the multifarious world is not *intrinsically* multifarious, simply because the multiplicity is *contingent* upon sensing.

To return to the question of "chemical reactions" versus "biological responses," and how the two are related, a contingentist view would suggest that the relation is the same as any smaller-level and larger-level phenomenon. Chemical reactions do not give rise to biological responses, and biological responses do not emerge from the assemblage and arrangement of chemical reactions. When biological responses are *viewed as* chemical reactions, this sometimes indicates a kind of theory that will be used in a limited context to make predictions—for example, the prediction that a particular kind of protein is necessary for a cell's movement in the presence of sugar. It also sometimes indicates a kind of slippage, where chemical reactions are seen as more *basic* than biological responses because they are seen to have the power to cause biological responses, instead of chemical reactions simply being regular events that come into view every time biological responses are dissected. Again, from a contingentist point of view, the idea that the smaller level could be more basic or fundamental than a larger level simply does not arise. This is a stronger claim than a kind of tolerant pluralism that says, "All phenomena are worthy of examination at their own scale and in their own way (though the smallest scale is still all that exists ultimately)." The contingentist view is that *no* scale or level of organization exists intrinsically, and that scales or levels of organization only arise contingent upon conceptions of relations "between" levels.

The idea of a "biological response" in contrast to (mere) "chemical reactions" is quite interesting because it is where the idea of *agency* arises in biological accounts. Agency is considered a kind of "larger-scale" or higher-level phenomenon, contrasted with the smaller-scale and deterministic chemical reactions. Agency is the sense that an entity can make something happen in the world without *itself* being made to happen.[34] We say a person has agency when she freely chooses, for example, the direction in which she will place her next step, where "freely chooses" means that she was not somehow *made* (coerced, influenced, determined) to take that step in that direction. Agency is thus a kind of causal account—and, as is now familiar, a contingentist view of causal accounts does not involve anything *making* anything happen in the world. In this view, billiard balls do not make things happen in the world any more than cells or people do. Again, billiard balls are discerned as discrete entities and cognitively linked with other entities. When this billiard ball strikes, that billiard ball moves, but no intrinsically existent causal power links the two. Biological agents operate similarly, themselves having to be discerned as discrete entities that are placed within causal accounts and are said to have the power to cause events.

If we say that biological responses are chemical reactions, the following intuition commonly arises: "Physical objects are determined and have no choices. Because living beings are physical, they are determined and without choice." For contingentists, this claim is dubious on at least two grounds:

1. Objects are not themselves self-presented or determined as objects. No object exists intrinsically. Objects are bounded and held continuous by cognition, and are narrated together in causal accounts. We do not know in advance what will be experienced as an object and what will not. Therefore, objects are themselves indeterminate in this sense.

2. Free choices are not somehow themselves uncaused, but are themselves always dependent. The "free agent" (the organism, living being, biological system, or conscious agent) arises only dependent upon countless factors, as will be explored more deeply in the next chapter. Though these dependencies are sometimes described as "constraints upon choice," they can equally be described as "what makes choice possible at all." This matter will also be discussed in more detail in Chapter 4.

The question of determinism versus agency (or fate versus free will) has held Western philosophy in thrall for centuries, but this does not mean that biology needs to somehow answer that problem—or rather, it does not mean that biology even has to accept it *as* a problem. Biologists *have* taken on the problem or seen it as a compelling one,[35] but this is not a *necessary* endeavor. The problem of determinism versus agency does not arise *as* a problem within a contingentist framework, just because neither agents nor causal powers are taken to be intrinsically existent from the start. If agency and causation have a place in contingentist accounts, it is as a pragmatic matter of being able to give more or less satisfying causal accounts, not as a self-presented problem of how and where agency has arisen from an otherwise deterministic universe.

Biological subjects are not intrinsically existent agents, but neither are we somehow *deceived* in thinking that the subject has agency when the subject is *really* (or nothing more than) a set of molecules interacting in deterministic ways. Both subjects-as-agents and interacting-molecules-as-agents arise contingent upon being bounded as subjects and placed into causal accounts. The subject is not the same as molecules; neither are molecules somehow more basic than the subject. The notion of the subject will itself be explored further in the next chapter.

Finally, the sense of an ambiguous and perhaps irreconcilable relationship between the physical and psychological may arise in part because the metaphors of the physical and the psychological are, from the first, defined in contrast with one another. Consider the following connotations of the physical versus the psychological:

Physical phenomena	Psychological phenomena
Dumb	Intelligent
Passive	Active
Manipulated	Manipulating
Unfeeling	Feeling
Inanimate	Animate
Nonsentient	Sentient
Determinate	Indeterminate (free)
Static	Dynamic
Choiceless	Choosing
Dull	Vibrant
Cold	Warm

The idea of matter, or of physical phenomena, is itself a metaphor whose resonance and power derives from its contrast with life and mind. There is nothing intrinsic to matter that makes it passive, dull, cold, and determined. Yet the intuition that a strong divide between physical and psychological phenomena needs to be maintained remains. Anxieties from multiple directions fuel the need for this divide. On one hand arises an anxiety that a world that is fundamentally physical is a fundamentally dead, determinate, and mechanical place. On the other hand arises an anxiety that a world that is fundamentally psychological is a fundamentally indeterminate and therefore a mystical, supernatural, and inexplicable place. However, erecting one realm to defend against the other has only strengthened both, such that we are left believing that these are ultimately our only choices in answering the question, "What *kind* of place is this world, anyway?"

All of this suggests a realm in which biology—having borrowed language from physics and chemistry and from psychology—may create its own unique metaphors. If a living organism is not quite like the physical idea of an assemblage of chemical reactions, and not quite like the psychological idea of the unified behaving agent, then what else is it like? Because physical and psychological metaphors depend upon one another and operate in opposition to one another, there are various ways in our language of relating them with the use of various prepositions. Life is considered what is either *both* physical and psychological, *in between* the physical and the psychological, *either* physical *or* psychological, or transformed *from* the physical *into* the psychological (or vice versa). Another metaphor for life may also be that which is *beyond* the physical and psychological altogether. By "beyond," what is suggested is not the sense of somehow "independent of," but simply the sense that biological phenomena do not have to operate only in relation to this dichotomy, even if to mediate between them. Just as the metaphors available for describing the phenomenon called "a cell" are not exhausted by its comparisons to a kind of animal, thermostat, chemical reaction, or standing wave, similarly, the metaphors available for describing living systems generally are not exhausted by the comparisons to passive materials or active agents. This suggests an enlivening view of phenomena, so to speak, one that does not buy the vibrancy of one side of the animate-inanimate divide with the deadening of the other. Neither does it entail a panpsychic[36] view that imbues all physical particles or forces with consciousness or an intrinsic vitality.

In sum, the signal transduction concept operates at a fascinating nexus where chemical reactions either become or give rise to biological responses. Physical, determinate, lower-level reactions are said to cause psychological or vital, agent-driven, higher-level responses. Yet, from a contingentist view, one cannot cause the other by virtue of a causal power. Moreover, the two "types" of phenomena are defined in opposition to one another from the start, yet there are other interesting and potentially useful ways of considering them, as will be discussed more deeply in the next chapter.

REVIEWING SENSING: NEW VIEWS OF TRANSFORMATION AND CHANGE

Before the idea of signal transduction (that is, for the proto- or early pharmacologists), the cell (or protoplasmic molecule) came into focus *as* reactive substance—perhaps *more* reactive than other complex chemicals. For the microbiologists, the cell came into focus as an organism, a kind of animal. The latter group clearly recognized the cell as a unified entity, a behaving and responsive whole, and a kind of agent. Though the extent of that agency could be (and continues to be) hotly debated, the *idea* of agency still came into focus as a point worthy of debate. If the cell is *like* an animal (maybe even like a human animal), and if the animal has qualities like behaviors, responses, senses, and agency, however expansive or limited, then the cell as understood by the microbiologist has all of these qualities as well.

This tradition of viewing the cell as a kind of animal-agent continues among microbiologists studying the lives and diversities of single-celled organisms. It also continues among systems biologists seeking the relation of lower-level reactions and higher-level responsiveness, and even among cognitive scientists examining the "simplest" forms of life in search of the "basic" features cognition. Though this cell-as-animal-agent view has its virtues, I would like to suggest that the earlier view of the cell as a kind of reactive substance also has several virtues, and is worth reinvoking and reworking today so that we may see behavior, sensing, responsiveness, reactivity, and agency from a different and useful angle.

The early cell physiologists and pharmacologists seemed to have little need to parse chemical reactions from biological responses and to explain how one gives rise to the other. For one thing, the cell had not been closely dissected into the hundreds of biochemical reactions that had been found through the dissection, so the cell was not exactly considered a *composite, aggregate*, or *interacting network* of reactions. The cell *as* the cell, the whole entity, was itself considered reactive. This suggests that the cell is not exactly reacting *to* an environment, but is reacting *with* an environment, as oxygen reacts *with* iron and where *both are transformed*.

One can conceive of the cell as an entity that changes in response to the environment, or conversely, one can conceive of a cell as an (active) organism that manipulates (passive) objects. In either case, the direction of change appears quite unilateral. Because we are ourselves the kinds of animals who tend to experience environments as what we respond to (as sensing, responding beings) and as what we manipulate (as agents), these kinds of unilateral causations seem intuitively to be what life is and how it unfolds for genuine organisms (that is, for those who come into focus at the level of organisms, and not as sets of biochemical reactions). We see flowers and pluck flowers, we touch apples and eat apples, and we experience ourselves as responding to and manipulating a world.

However, instead of only using our intuitive ideas about animals-as-agents to inform our ideas about cells-as-animals-as-agents, it can be helpful to run the shuttle the other way and use ideas about cells-as-reactive-substance to generate ideas about animals-as-reactive-substance. What comes to focus in this view is the way in which organisms and environments are transformed from moment to moment. For this is not simply the co-constructionist view that organisms and environments co-construct one another *sequentially*, as when an organism digests food from the environment and excretes wastes into the environment, which in turn change the organism's metabolism, and so forth.[37] Such a view still separates organism from environment and makes one the causal agent, and then the other, and so forth, so that agency shuttles between the two. The "reactivity" view suggests that organism and environment are transformed *simultaneously*, and at each moment.

Let us consider again ideas of change and stasis in the contingentist view, as examined in the previous discussions about sameness, difference, and energy. The transducer is not intrinsically the *same* transducer over time that changes form (or conformation). The organism is not intrinsically the *same* organism over time that is at one moment static and is at another moment undergoing a change in state (or a response). The signal is not intrinsically one object or energetic transference that is intrinsically unchanged itself though it induces changes in the transducer and in the organism. It is only when some aspect of the system is seen as the same thing over time that is changing (the transducer, the organism), and another is seen as the same thing over time that is unchanging (the sensed object, the signal, the external world) that we get the sense that the latter has produced a change in the former, and that too unilaterally. The view of cellular, or indeed, *organismal* reactivity as mutual transformation of substances can help us formulate a view that does not rely on either "transfers of energy" or "changes in objects":

> The organism, the sensing apparatus (like a transducer), and the external object that is sensed (signal, object, environment, or world) react with one another in every moment such that all are transformed or changed.

Such a view is not forced to rely on assumptions of the intrinsic existence of energy, sameness and difference over time, lower and higher levels, agency and determinism, or physical and psychological phenomena. However, some obvious problems remain. First, the organism, the transducer, and the external object are not intrinsically *substances* for all the reasons discussed in this chapter and the last. Second, they are not two intrinsically *different* things that are separate, and then meet, and then react. Third, neither are they intrinsically *transformed*, since this entails the very essentialism we were trying to dissolve:

that an object can change form, or be transformed, and be the same object. We may therefore try the following formulation instead:

> The organism, sensing apparatus, and external object do not interact with or react with one another (that is, they do not cause changes to each other by virtue of some intrinsically existent causal power). Neither do they remain the same objects that change (or are transformed) over time. Rather, they arise *new* in each instant.

In the moment of seeing an apple, you haven't just *discovered* or *met* the apple that has been there outside of you. Rather, in that moment, the apple arises as form, properties, continuity, vividness, and the like, for none of these exist intrinsically but arise contingent upon your activities (as discussed in Chapter 2). At the same moment, *you* are new, being not intrinsically the same thing before and after the seeing (as discussed earlier in this chapter). The apple is assimilated into your body, not via digestion, but via sensing. Your physiology is changed upon seeing the apple. However, this change does not involve the *transfer* of any substance—neither energy nor information or anything else, no matter how subtle. Though we might usually say, "The organism senses the apple" or "The apple caused a response in the body," we could also say, "When the apple is seen, the body is changed." And, yet another alternative: "When—depending on a human body—form, properties, continuity and vividness, arise *as* an apple, the body is changed." Furthermore, if we don't take the body to have some essence that somehow stays the same over time *despite* myriad changes in its form, instead of saying "The body is changed" we could equally say, "The body is *new*."

What do the standard views of sensing depend on? They depend on a separation between supposedly intrinsically existent subjects and supposedly intrinsically existent objects. They depend on strong delineations between what can become the organism ("assimilation") and what must remain separate from an organism (which "senses" an "external" world); hence, they depend on a divorce between the processes of metabolism and sensing. They depend on a notion of energy as a kind of substance, allowing the transfer of what is outside an organism to what is inside an organism, but while maintaining that the outside of the organism remains unchanged, and the organism remains the same organism. These views of sensing have also depended on and perpetuated the notion that lower and higher levels, agency and determinism, and physical and psychological phenomena are intrinsically existent, and are separate and interacting categories. A contingentist account of what is called sensing describes phenomena not as intrinsically existent, nor as interacting, nor as reactive, nor as changed or transformed over time. An alternative formulation that may come into view in the absence of the aforementioned concepts is that phenomena arise anew in each instant. Moreover, this "arising anew" occurs *dependently*—that is, phenomena bring each other newly into being in each instant. This view may become even more useful and pleasing in the next chapter, as we consider more closely the dynamism of organisms.

4

WHAT DO ORGANISMS DEPEND ON?
Bodies, Selves, and Internal Worlds

If objects do not exist inherently but arise dependent upon the cognitive activities of organisms, then who are these organisms? Who is it that is doing this work of aggregating and distinguishing and inferring? Who is it that relates to "what is" as objects and substances? If objects do not exist inherently, then one might assume that the organisms upon whom these objects depend must themselves exist inherently, for it is only they who can do the cognitive work necessary for objects to be experienced as objects.

In this chapter, I outline four common assumptions—commonsense beliefs or acquired anxieties—that strengthen the sense that organisms exist as intrinsically bounded and continuous *objects* (that is, entities that can be defined by an outside observer of organisms—such as a biologist, or someone visiting the zoo), and as *subjects* (that is, as beings who experience *themselves* as intrinsically bounded and continuous):

1. We assume that organisms have distinct spatial boundaries and temporal continuity. However, all of the problems that arise in attributing intrinsic boundedness and continuity to inanimate objects (as discussed in the previous chapters) arise in attributing this boundedness and continuity to living organisms. Moreover, organisms can seem even *more* bounded and continuous than objects in part because once they reach a certain stage of development, they appear to achieve a kind of "stasis." However, development never ends.

2. We assume that organisms are *selves*, in that they are (or we are) the coordinator of actions that exist independently of the experience of coordination, and/or we infer that they are (or we are) experiencers who exist independently of experience. However, the sense of there being an experiencer is itself an experience, and assuming that this experiencer is intrinsically existent leads to a regress ("Where does the experience of being an experiencer come from? Another experiencer?" and so forth).

3. We assume (or worry) that *not* taking other subjects ("other minds") to be intrinsically existent will make us vulnerable to thinking and behaving as if other beings were creations of our own minds, and not beings in their own right. However, we take one another seriously as subjects just because we depend on one another from the very first, not because we logically deduce that others exist intrinsically.

4. We assume that experiences arise from some fundamental basis that is *either* material (like a brain) *or* immaterial (like mind or consciousness). However, both are metaphysical assumptions that are not strictly necessary to account for the arising of phenomena.

ASSUMPTION OF THE BOUNDEDNESS AND CONTINUITY OF ORGANISMS

Organisms can be defined from a third-person perspective in a variety of ways, just as objects can be defined by an observer. But the same problems that arise when we try to justify the idea of the inherent existence of external *objects* arise when we try to justify the inherent existence of external *organisms*. Alternatively, we could say that, like an object, an organism is a system bounded in space and time. In this sense, the organism is *contingently* existent because (as discussed in previous chapters) only observers can bound phenomena in such a way.

In the course of daily life, it can be easy to take for granted just how extensive organisms are spatially, and how much is required to create the sense that an organism is *one thing changing* instead of *many events arising*.

Let's take it from the beginning. At the earliest stages of development, it can be quite difficult to define just where the organism begins and ends. A chicken embryo seen with the unaided eye is barely distinguishable from the yolk itself; it's just a small, darkish, almond-shaped spot (Figure 4.1a). At higher magnification, a small body comes into focus, but still indistinct in its way (Figure 4.1b). One part of the body is well defined (a head and top part of the back bone on the right side of image), even while the rest of the body hasn't quite formed, and remains a largely undifferentiated mass of cells (left side of image). Figure 4.1c depicts a three-day-old chicken embryo, photographed through a small break in the eggshell. Blood flows through blood vessels from the shell to the yolk to the embryonic heart (at this point, just a simple, pumping tube), and back out to the blood vessels on the yolk. The blood is carried out toward the shell, where carbon dioxide can diffuse out of the blood and oxygen can diffuse in and be carried back to the embryo. The shell, in a sense, thus functions as the organism's lungs at this stage.

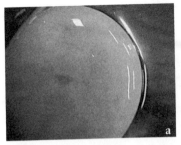

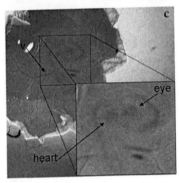

FIGURE 4.1. Chicken development. (a) Egg yolk with three-day-old chicken embryo. (b) Day-old embryo, magnified. (c) Three-day-old embryo, imaged through a tiny break in the shell. Images courtesy of: (a) Mickey A. Latour, (b) Carolina Biological Supply Company, and (c) Anjul Davis.

The process of early development makes vivid the difficulty of bounding "the organism" given the seamless continuity of "embryo" and "environment." Even for developmental biologists who are habituated to observing early development, the way in which the embryo seems to slowly emerge, to gradually take form from a formless mass, can continue to astonish and perplex. Each of us came about through this process, the process of *becoming* a kind of foreground distinct from a background. It was not always a given that we would be the stage actors, and the environment relegated to the role of theater. Though it's now an obscure time before remembering and even beyond imagination, there was a time when we were literally nowhere to be found. Now that it's clear to ourselves and most everyone around us that we're decidedly *here* (that is, bounded, continuous, and easily found), living seems to be a matter of maintenance rather than creation. By eating and breathing, we seem to keep a vehicle running that was manufactured long ago.

For many, breathing is relatively effortless and automatic, strengthening the sense that we have largely been made and do not have to do much to keep this body going. We can breathe voluntarily, but we also breathe automatically, without consciously being aware of the breathing or *deciding* to breathe. Indeed, if a person stops breathing for a period, he will automatically gasp for air. In the absence of oxygen, carbon dioxide accumulates in the blood, creating carbonic acid. Receptors in blood vessels change in this increased acidity, and stimulate the nervous system, causing the person to gasp whether he wants to or not, whether conscious or not. For someone surrounded by air at the time, this gasp reflex is a helpful thing. For someone who happens to be underwater at that time, however, the gasp will draw water into the lungs, completely depriving the body and brain of oxygen. That's death by drowning.

Staying alive depends on a vast number of details coming together moment after moment after moment—small details that are largely taken for granted, like whether or not one's head is immersed in air (and not water) when one's lungs gasp reflexively. It is not necessarily the case that a body is created once and is just maintained thereafter. So much has to come together to bring a body into being at every moment—as much in this moment as in the moment of conception or the moment of birth. The body is always being created.

Conceiving of organisms as changing objects or as bounded but open systems is a very useful metaphor for certain purposes. It's a tool, like a picture frame, that allows us to appreciate and study particular compositions. There is also a larger set of metaphors available to us. Maybe an organism is like a party. The party metaphor, in contrast to the object metaphor, places its emphasis on the way in which an organism (and indeed, each and every moment of development) is an event that happens once and is never repeated. Nonetheless, we find commonalities with other events, and we can say, "I've seen many parties, and they share these common features." What makes a party is the coming together of many factors, over and over. When those factors don't come together at any given moment—the DJ makes a poor choice, the cops come knocking, the punch is all gone—then the party's over.

Organismal development is quite like that. A young turkey ingests gravel. The gravel settles in the gizzard, where the turning of these small rocks helps to crush nuts that the turkey eats so that they become digestible.[1] The turkey is not born carrying around everything it needs to live in a personal knapsack called its body. The turkey comes together over time. The turkey, as writer Elbert Hubbard famously said of life, "is one damn thing after the next."[2] Faced with a turkey, one could ask, "How does this thing work?" One could equally ask, "What is happening?"

To summarize, organismal development is the process by which an organism comes into being. Early development makes vivid the ways in which living bodies come into being radically dependent upon nonliving matter. In early development, one can easily glimpse the seamlessness of inanimate and animate matter. Moreover, development doesn't end. At no moment does a living body cease to depend radically upon nonliving matter, no matter how much we may be in the habit of experiencing our own bodies and seeing other bodies as radically distinct from mere "surrounds." Just as a very young organism is not neatly bounded (for example, the embryonic chick whose egg shell acts as the embryo's lungs), neither is the mature organism (for example, the adult human being whose lungs are surfaces exposed to the "outside" world, and depend critically on that world being air and not water at every moment).

ASSUMPTION OF THE COORDINATOR AND THE EXPERIENCER

Unlike objects, some organisms, like humans, can define *themselves* as organisms from a first-person perspective.[3] That is, the body can be organized into an object by an "internal" observer. This first-person observer is what could be called the self or subject. The subject would need to group its various parts into one thing, locate itself and bound itself as separate from a surrounding space, and stitch its moment-to-moment existence together into a single continuous experience.

We are still left with the question of who is doing all of this cognitive work. In *The Unity of Consciousness*, Alex Cleeremans writes,

> From a first-person perspective, information processing does indeed seem to take place on some sort of Cartesian theatre where different experiences are neatly integrated as part of a single play. Yet, when one explores how the brain actually processes information, there is little unity to be found. Discrete regions of the cortex process separate streams of perceptual information, and often asynchronously so. Likewise, while the brain does contain many multi-sensory convergence zones, no region can appropriately play the role of the sort of "ultimate" convergence zone that the Cartesian theatre metaphor seems to mandate the existence of.[4]

Cleeremans suggests that the first-person intuition about oneself as a unified subject experiencing many things simultaneously warrants a search for an integrator, a center where various stimuli are organized into a single stream of experience. The search for how the human brain functions as a whole is often articulated as a search for this integrator.

The question of whether consciousness is unified is a longstanding philosophical problem,[5] as is the question of "who" controls awareness. Is there a self who exists independently of the thoughts, feelings, and behaviors and has some control over them? The self is assumed to be above or separate from the thoughts, emotions, and actions, a kind of chooser, decider, or integrator. But what is the self *other* than thoughts, feelings, and actions? Who is the chooser behind the choice, or the thinker behind the thought? Who do thoughts *belong* to? This brings us to that million-dollar question we've each encountered at some point in our lives: Exactly who is in charge here?

Again, as Garfield writes, "Searching for the tree that is independent of and which is the bearer of its parts, we come up empty." Similarly, searching for the self that is independent of and which is the bearer of its parts—limbs and organs, yes, but also moods, personality traits, feelings, and thoughts—we come up empty. When seen as something independent of thoughts, the self becomes a kind of "ghostly level"—similar to the essential "tree" separate from the parts of the tree or an essential objects separate from the properties of an object.

The self could also be an observed coordination between thought and an action: "I want to move that cup. My hand is moving toward the cup and I'm picking it up." Like any other causal relation, this relation is not explanatory, but descriptive (see Chapter 2, on assumption of intrinsically existing causes). That is, a thought arises, then the experience of moving a hand arises, and then the experience of seeing and feeling a cup lifted by the hand arises. There is a feeling that the hand moved and the cup moved *because* of the self's agency: "This happened *because* that happened," instead of simply "this happened, then that happened." The sense of self can be evoked as an explanation for a set of regularities, but it does not necessarily exist independently of these regularities. That is, the self is not an intrinsically existing thing. (I explore this idea of "self-as-agent" further in Chapter 5 in the section "On cognitive patterns and cognitive dissonance".)

A well-known psychological experiment makes vivid how powerfully a sense of coordination can give rise to a sense of self. Commonly known as "The Rubber Hand Illusion,"[6] the experimental setup involves a subject sitting across from a dummy hand. The subject's own hands are under the table, away from view. One experimentalist uses a fine brush to stroke the dummy hand while another experimentalist uses a fine brush to stroke the subject's hand under the table. The subject, seeing the dummy hand being stroked and feeling their own hand being stroked in the same manner and timing has a vivid sensation that the dummy hand *is* their own hand. That is, the subject feels that his or her hand is being stroked, but that the hand that is feeling the stroking is not under the table, but is sitting across from them—a plastic dummy hand. This experiment and several others like it demonstrate how things that are not usually experienced as part of the self can *become* experienced as the self.

To give a hypothetical example, imagine that if every single time you thought, "It's time to turn the page of this book," the page turned (either by itself, by the movement of the wind, or by someone standing over your shoulder and turning it in such a way that their body is hidden). You might get the distinct and perhaps eerie feeling that *you* had turned the page, or at least that you—the subject, the self—had caused the page to be turned. The sense of coordinated action is necessary to give rise to such an experience, though an intrinsically existent coordinator is not.

Who is it that claims to be a self? The statement "I believe that there is no intrinsically existing self" can seem self-contradictory, but it needn't be. "I" is a linguistic device used in communication and social interactions, just as proper names are such linguistic devices.[7] "I" or "self" function in language as terms with referents; however, referents have no inherent existence. The "subject," the "I," is a term with no inherently existent referent, just like the term "flower" has no inherently existent referent (again, what is called "flower" is not *by itself* bounded or continuous). Again, to revisit Rorty's comment on how language,

> Perhaps saying things is not always saying how things are. Perhaps saying *that* is itself not a case of saying how things are. . . . We have to drop the notion of correspondence for sentences as well as for thoughts, and see sentences as connected with other sentences rather than with the world.[8]

Just as one could "drop the notion of correspondence" between sentences and the referenced world, similarly, one could "drop the notion of correspondence" between sentences and the speaking subject.

There is not necessarily an experiencer who is separate from, simply, experience. What, then, explains differences in the content of experiences? I have obviously experienced different things from you today, since I have been in a car along a particular highway while you were, say, at home sorting the mail. A great deal explains differences in the content of experiences: bodies, objects, memory, language, perception, thought. But none of these explanatory categories implies intrinsically existing differences between subjects, that is, between experiencers. My apprehension of the steering wheel depends on the steering wheel being there. The steering wheel being there depends on my relating with it as a steering wheel—cognitively bounding it, turning it, knowing that it is definitely a "steering wheel" and not an "air conditioner dial" because turning it shifts the car and doesn't modulate the A/C. Your not apprehending the steering wheel depends upon the steering wheel not being before you. The steering wheel not being before you depends upon your experiencing whatever *is* before you as "not a steering wheel." So, yes, we have different experiences. Those experiences differ based on circumstances (objects, situations), and those circumstances themselves depend upon experience (knowing, doing).

It is the phrase "to *have* an experience" that is potentially misleading. It resembles the phrase "to *have* properties," or "to *have* a self," as explored above. What can *have* properties? What thing is separate from and the possessor of its properties? And what can have a self? Is a self separable into the possessor and the possessed? What self can then *possess* a self? Similarly, what experiencer can *possess* that experience? Experiences aren't intrinsically *yours* or *mine*. The idea that experiences are the kinds of things that can be possessed is the product of linguistic shorthand, which, taken too literally, gives rise to the sense of a separation between experiencer and experience. What can be coherently said about experiences without positing a metaphysical, ghostly, or inherently existent experiencer, is simply that experiences *arise*—full stop.

ASSUMPTION OF INTRINSICALLY EXISTENT "OTHER MINDS"

The idea that subjects may not exist inherently may give rise to the following worry: "Sympathy, empathy, and indeed, all moral relations with others require an understanding that others exist, and that they experience in ways that are independent of our own experience. How can we have empathic relations without assuming the intrinsic existence of other subjects?"

Philosopher Alva Noë has skillfully addressed this concern in his book *Out of Our Heads* by noting that human beings can be seen as born already bound in a social world and therefore in empathic relations, in the absence of a theoretical assumption that subjects (other selves, other minds) exist intrinsically:

> The point that I want to make is that the question of whether a person is in fact a conscious person is always a moral question before it is a question about our justification to believe—even if it is also a question about our justification to believe. Even to raise the question of whether a person or a thing has a mind is to call one's relation to that person into question. And this is the point. For most of us, most of the time, our relations to each other simply rule out the possibility of asking the question. For the question can only be asked from a detached perspective that is incompatible with the more intimate, engaged perspective that we actually take up to each other.

Noë's account suggests that whatever it is that saves us from relating with one another as unconscious automatons, it is not a theoretically sound defense of the intrinsic existence of Other Minds. Such a defense doesn't, well, exist. Nothing proves beyond doubt that other humans are conscious and feeling; as Noë says, "Our commitment to the minds of others is not based on evidence." The commitment therefore stems from other quarters, another kind of logic, reason, or empiricism called, simply, everyday life.

We are right to take one another seriously as experiencers. You exist as I exist:

contingently. We do not have to take one another as intrinsically existent experiencers to know that most everyone has a very strong sense of being an experiencer in an intrinsically existent way—that is, of feeling like the possessors or at least the sources of experiences, of joys and sorrows, memories and ambitions. Knowing what it's like to experience oneself as an experiencer, of course we would feel empathy with those who themselves are experiencing the same.

If, as Noë suggests, our mutual constitution is vivid at birth and in infancy, it is also quite vivid at death. Philosopher Judith Butler writes of the psychologically profound ways in which grief reveals our relations with one another as not merely "interactive" but as properly *constitutive* of identity itself:

> When we lose certain people, or when we are dispossessed from a place, or a community, we may simply feel that we are undergoing something temporary, that mourning will be over and some restoration of prior order will be achieved. But maybe when we undergo what we do, something about who we are is revealed, something that delineates the ties we have to others, that shows us that these ties constitute what we are, ties or bonds that compose us. It is not as if an "I" exists independently over here and then simply loses a "you" over there, especially if the attachment to "you" is part of what composes who "I" am. If I lose you, under these conditions, then I not only mourn the loss, but I become inscrutable to myself. Who "am" I, without you? When we lose some of these ties by which we are constituted, we do not know who we are or what to do. On one level, I think I have lost "you" only to discover that "I" have gone missing as well. . . .
>
> What grief displays is the thrall in which our relations with others holds us, in ways that we cannot always recount or explain, in ways that often interrupt the self-conscious account of ourselves we might try to provide, in ways that challenge the very notion of ourselves as autonomous and in control. . . .
>
> Let's face it. We're undone by each other. And if we're not, we're missing something.[9]

One narrative of grief is that one day we will "go back to feeling like our old selves again." But the experience of grief can also reveal that after the loss of a loved one, one is not the same. "A contingentist account," so to speak, of grief and its healing might not advocate for the return to an intrinsically existent self, but for a recognition that, yes, the self will now have to be made anew—as it always has been. After loss, one finds oneself as a new self in a new world in which the loved one is no longer present to hold in place the things that they held in place for the old self, and so the new self can hold together in new ways.

The experience of being a radically independent self only arises when everything is going fine—that is, when everything upon which our sense of independence *depends* is reliably present. In conditions of good fortune, the smell of our critical resources never reaches our nostrils. However, in the absence of loved ones and other critical deprivations (safety, food, and so forth), the subject does not cohere along the same lines as it once did—thus vividly revealing, yes, one's contingent existence.

ASSUMPTION OF A GROUND: PHYSICALISM, IDEALISM, DUALISM, AND CONTINGENTISM

Even if one agrees that the self cannot be some coordinator, integrator, or bearer of experience that is independent of those experiences, experiences keep coming—so we think they must come from some*where*. The self could be seen as, if not the *bearer* of experiences, then at least the *source* of experiences. Some neurobiologists are even looking for the self as a place in the brain which is not an integrating center, but a place which gives rise to the distinct feeling that there *is* an integrating center—that is, a physical place where the *idea* or experience of the self comes from.[10]

In order to discuss this assumption, it is helpful to contrast contingentism with three other perspectives: dualism, physicalist monism, and idealist monism. The first two of these—dualism and physicalist monism—are probably the most popular. These perspectives are typically characterized by their orientations toward the nature of reality; but even more interesting may be their orientations toward mystery. I give an overview of each position in the following list and outline them in more detail in Table 4.1. These overviews are brief, but not cursory. Monism(s) and dualism are among the most thoroughly considered and debated topics in several philosophical systems. The purpose of invoking them here is to outline some of their basic features, thus offering a useful clarification of the contingentist view.

1. Dualists claim that some phenomena are physical, and some are nonphysical (nonmaterial, metaphysical, or perhaps spiritual). Physical phenomena can be understood, explained, and predicted in physical terms, but, dualists believe, a number of phenomena cannot be understood in physical terms. For those, dualists evoke a separate sphere, substance, dimension, or process. The separate "thing" could be experience, mind, consciousness, spirit, soul, divinity, etc. As far as I can tell, dualism is a very popular worldview.

2. Physicalist monists (abbreviated as physicalists herein) claim that all phenomena are fundamentally physical and can be understood in terms of physical laws. "Physical" means matter—fundamental particles—but also includes fundamental forces, energy, space, and time. These would all be considered phys-

ical phenomena because, physicalists believe, they can be interrelated by physical laws. Phenomena that cannot currently be predicted by physical laws are not *in principle* impossible to predict with these laws—they are simply too complex, say physicalists, for our current (and perhaps our future) cognitive capacities and technologies. Experience, mind, consciousness, and/or the paranormal can all in principle, they maintain, be predicted by physical laws, and it is not impossible that they will be some day. Many scientists or admirers of science are physicalists.

3. Idealist monists (abbreviated as idealists) claim that all phenomena are fundamentally mental. "Matter," "physical substance," "space," "time," and "physical laws" are all experiences that arise in consciousness. We know they arise in human consciousness, and might have varied opinions about whether they arise in nonhuman consciousnesses. Consciousness imagines the existence of a substance that is separate from itself, called matter, but in reality, everything is consciousness. Consciousness is itself a very subtle kind of substance which gives rise to all phenomena. All phenomena, whether according with physical laws or not, can be explained as mental, or perhaps subjective. Thoroughgoing idealists are probably quite rare.

4. Contingentists can be seen as a species of skeptics or ontological agnostics. That is, they take a position *against* dualism, but not *for* monism of any kind. Contingentists are not physicalists because they see the assumption that physical substance is fundamentally what exists as an unjustified assumption, since physical substance is itself a kind of experience just like any other. In this sense, they agree with idealist monists. However, they are not idealist monists because positing the existence of single "substance" (consciousness, subjectivity, or even moments) is also an assumption. Contingentists make neither assumption.

Each of these perspectives produces a different answer to the question, "Where *do* experiences come from?" The physicalist monist says experiences arise from physical substance (objects, matter). However, as argued before, physical substance or objects or matter are *themselves* experiences—the experience of "physical substance," "objects," or "matter." The idealist monist says that experiences arise from some very subtle substance, perhaps called mind or consciousness. However, similarly, the thought of that subtle substance is itself an experience.

The contingentist says that experiences arise like everything else: contingently. In the absence of living organisms, experiences do not arise. In that sense, experiences are dependent on bodies and lives and do not somehow exist inherently as idealists might claim. In the absence of physical substance, living organisms do not arise. Yet, contra the claims of physicalists, contingentists say that in the absence of living organisms, and the experiences of living

TABLE 4.1. Comparing dualism, physicalist monism, idealist monism, and contingentism

	Dualism	Physicalism	Idealism	Contingentism
Ontology: What is? What exists, what is fundamental, what is real?	Some phenomena are physical, and others are nonphysical.	All phenomena are physical (particles, forces, energy, time, space).	All phenomena are mental (thoughts, awareness, consciousness).	Phenomena are neither physical nor mental.
Epistemology: How are phenomena known? Explained?	Physical laws explain physical phenomena. Nonphysical phenomena are either fundamentally inexplicable or explicable through nonphysical laws.	Physical laws explain physical phenomena, and all phenomena are physical.	Consciousness explains the existence of phenomena. Both the knower and the thing known arise from consciousness.	One may think that we can explain phenomena, but we can actually only describe relationships between phenomena.
Assumptions: What must we take as a given, a basic assumption that can't be defended by reason or empiricism?	Physical substance exists separately from consciousness that perceives it, yet the two somehow interact.	Physical substance is fundamental, and consciousness itself is a physical process.	Consciousness exists, even though since everything is consciousness, nothing outside of consciousness can vouch for its existence.	No concept can be assumed or affirmed. All positions regarding the intrinsic existence of things are indefensible.
Mystery: What is mysterious? How do we handle mystery?	Whatever cannot be explained by physical laws can either be explained by nonphysical laws or is fundamentally inexplicable	Whatever cannot currently be explained by physical laws can in principle be explained by physical laws at some future time.	Consciousness explains everything. Mystery is simply a mental construct like everything else. Physical laws are patterns that are perceived as such by consciousness; the inexplicable is just the absence of the pattern.	As long as we try to explain the ultimate or intrinsic existence of things, mystery will appear real, and we will relate to it either as something to be conquered or something to be held sacrosanct.

organisms, physical substance *itself* does not arise. That is, again, matter, objects, and physical substantiality do not necessarily exist inherently—objectness and substantiality are crucially *experiences*. Whatever is experienced as an object has no inherent property of objectness, materiality, or substantiality. To put it all together succinctly in contingentist terms, physical substance is necessary for organisms, which are necessary for experiences, which are necessary for physical substance. Or, to start the circle in some other place, experiences are necessary for physical substance, which are necessary for organisms, which are necessary for experiences.[11]

What we have is what mathematician Douglas Hofstadter calls "a strange loop."[12] As the name implies, it's rather puzzling. However, it is not an incoherent or self-contradictory position. The contingentist account does not seem to make sense if we are trying to find *the* one original cause of experiences—the fundamental basis of experiences, the place where we can say experiences are somehow "grounded" or "located." The monists (physicalists and idealists) and the dualists provide some kind of ground. Physicalists say the origin of all existence is physical (i.e., the experiencer is an arrangement of matter, and objects of experience are arrangements of matter), and idealists say the origin of existence is the subtle substance of mind, the subject. If mind and matter are fully mutually entailing, constantly giving rise to each other, then *must* one come first, or could it? The "strange loop" account is as well-composed and satisfying as Escher's *Drawing Hands* (Figure 4.2), even though we have no reason to assume that the top hand, the bottom hand, the pencils, or the paper came first.[13]

In sum, experiences arise, but this does not justify the assumption that there is an intrinsically existing experiencer who is the source of those experiences—either a brain or body independent of experiences that "secretes" experiences, or a consciousness or subject independent of experiences. Even though experiences depend on bodies and worlds, this does not mean that bodies and worlds are themselves independent of those experiences and "producing" the experiences instead of the other way around.

WHAT DOES YOUR LIFE DEPEND ON?

What does your life depend on right now, in this instant? What are all the things that are happening right now such that you don't collapse dead? Here is my list, quickly jotted:

> blood, breath, air, gravity; the blood keeps moving; the heart keeps beating; the sun has not gone out; the air is not polluted for this moment; my head is immersed in air and not water; nothing is crushing my bones under its

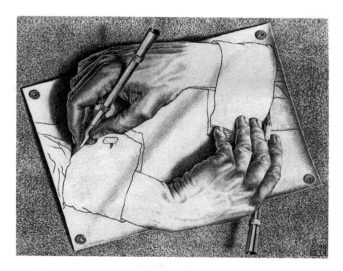

FIGURE 4.2. M.C. Escher's *Drawing Hands*. © 2015 The M.C. Escher Company-The Netherlands. All rights reserved. www.mcescher.com. 1948, Lithograph, 28 × 33 cm.

weight, neither car nor meteor; no bullet nor claw has crossed my flesh; nerves fire; there are reserves of energy in my body—it is not exhausted, it has the thrust to continue; my environment is neither freezing nor burning; this planet maintains respectful distance from the sun.

And what historical conditions does your life depend on? What are all the things that happened to go one way rather than another in the past such that you are alive right now? Again, my list reads:

my parents conceived me—they were humans and not cats; people held me and spoke to me when I was an infant; people treated me as if I mattered; I loved my family enough to decide to live when I was in pain; primates formed social groups; cities came together and ensured that I had clean water to drink every day; when I drove, the parts of my car functioned well enough that it didn't steer out of control and crash at a high speed, and the intricate parts of the other cars hummed along the same way; no driver's heart gave out, nor did the baby in the backseat put something in her mouth and turn blue, panicking the driver and causing him to swerve; food has consistently found its way into my esophagus and not my windpipe; matter had a tendency to be lumpy and form aggregates; and I didn't die, I didn't die.

These lists, of course, could be books in their own right. What is often omitted from such lists, however, is self-perception itself. Given the vast set of conditions upon which your life has depended historically, and the vast set of conditions upon which your life depends at this very instant, what is it that makes your life *your* life? If life arises at the nexus of such innumerable and far-flung factors, what makes the experience of life such a personal affair?

It is true and widely acknowledged that each life depends on specific material and energy flows: Water, for example, does have to enter and exit bodies at a certain rate in order for these bodies to live. Yet, in order for life to be "*a* life" or "*my* life," something more than material or energy flows are required. *Experience* is required. It takes *experience* to distinguish "a living being" from "its surrounds" or from "other living beings." In other words, though material and energy flows are necessary for life, it is the act of distinguishing that is necessary for *personalizing* life. Life and nonlife have no intrinsic existence; in the absence of distinguishing cognition, life and nonlife are not necessarily distinct. So what does my life depend on? It depends on all of the material coincidences outlined in the preceding lists—and it depends critically on the experience of "*my* life" as distinct from "life," or just "this."

Again, where do experiences like "*my* life" come from? A biologist friend and colleague once thoughtfully admitted to me, "You know, whenever I think about perception and ask myself how it is that I can perceive things and think of things, I always visualize my own brain doing something. I find it easy to forget that *that* visualization of my own brain is also a concept." The experiences "my body," "my life," "my brain," or "my experience" themselves depend upon "my body"—which is itself an experience. We use sentences like, "My brain is generating the experience that my brain is generating the experience," and "My brain is generating the experience that 'brain' is a thing that is separate from 'experience.'" Though one might commonly go through the day locating experiences in a particular place ("body," "brain") and attaching them to a possessor ("I," "my experiences"), these formulations illustrate the awkwardness of such an approach, and the comparative simplicity of the phrase, "experiences arise." This simple phrase does not unnecessarily add a location or a possessor to the phenomenon. This is how the *subject* can disappear, even if the organism as a nexus outlined for various purposes does not.

———

What do organisms depend on? They depend on delineations of distinct spatial boundaries and temporal continuity. The organism-as-subject depends on the sense that experiences arise from a locus, whether material or immaterial. Finally, whereas the organism-as-object depends on material and energy flows, the organism-as-subject depends additionally on the *experience* of the nexus of such flows as a separate, intrinsically existent thing.

would do it but it felt very strange to him, very bizarre, as if he were violating some law of physics, or something like that. It gave him gooseflesh or he would shudder and get frightened. It's very real.[1]

Though nothing physical can obstruct a phantom limb, Livingston's patient experienced feelings of fear and confusion upon experiencing the limb as violating physical laws. This feeling of revulsion, anxiety, or aversion when faced with a thought or experience that doesn't fit with one's most deeply held beliefs and expectations is called "cognitive dissonance."[2] When dreaming, one dreams a world that has no physical substance,[3] yet still often behaves according to very ordinary physical laws. When our dreamed selves violate those "physical" laws, we can sometimes experience cognitive dissonance.

To give a personal example, I once had a dream in which I found myself in a large house. I suddenly became aware that I was dreaming (this phenomenon is called "lucid dreaming"). Looking out of the windows of the house, it looked like I was on a floor of the house just a few feet from the ground. The moment that I realized that I was dreaming, I decided to take advantage of dream life and experience flying. My dreamed self ran straight to a door at the end of a hallway. Instead of being on the first floor of the house, I found that I was on a balcony hundreds of feet above the ground. My dreamed self climbed onto a ledge, spread out its arms, and was about to jump, but hesitated out of fear. I then thought, "It's really strange that I'm afraid right now. After all, I know this is a dream, and I won't get hurt." Repeating the statement "it's just a dream—I won't get hurt" a few times, my dreamed self gained the courage to jump. I spread out my arms and took the jump, then noticed that I wasn't flying, but instead falling. I couldn't *make* myself fly—I could only float to the ground. I questioned my agency, wondering why I couldn't control my movements and my circumstances even though this was *my* dream.[4] Moreover, I noticed that though I wasn't wearing a skirt at the moment of jumping, as soon as I started falling, I was wearing a huge, multilayered skirt that billowed out like a parachute, making for a slow, gentle fall. Noticing this, I thought, "My mind must have created this skirt in order to try and account for the fact of my gentle fall." Then, when I landed gently on the ground, I saw a beautiful, architectural, ordered city around me. Upon noticing this order, I thought, "My mind is taking what it knows as patterns called 'physical laws' and creates an ordered world, like this one."

Several features of this lucid dream suggest how a mind while dreaming can create a seemingly substantial and seemingly external world, and go through cycles of being by turns constrained by the limits of that world and freed from those limits. I draw attention to a number of important moments of awareness that change the dream from constraining to free and back again:

1. Noticing the absence of external constraint: "This is just a dream. It's not the same as waking life—the rules of the waking world don't apply." This is the key moment that creates a sense of freedom and possibility within a dreamed world. It comes from a dreamer being aware that he's not *really* the dreamed self, that the dreamed self can't *really* get hurt, and that dreamed physical laws don't *really* exist, and thus can't really constrain. "Really," of course, means "real from the perspective of waking life." Lucid dreaming brings some of the perspective of waking life into dreaming life.

2. Desire and the exercise of will (that is, feeling agency): "Given freedom and possibility, I want to do something with it. I want to fly. If I wake up, I'll lose my chance. I'll move my (dream) body in a way that allows it to fly." The dreamed body feels somewhat responsive to a sensation normally called "will." In a normal dream, the dreamed body may feel like it's acting according to its own will, or it feels that it has no will and no control over the dreamed body or its circumstances. In a lucid dream, the dreamed body feels like it's acting according to the will of a dreamer. There is some awareness of relationship between the dreamer and the dreamed that doesn't arise in a normal dream. This is because in a normal dream, there is no awareness of the existence of a dreamer—the dreamed self and dreamed world are taken to be all that is. Hence the instant of surprise when the dreamer wakes up from a normal dream and realizes that the dreamed self and the dreamed world were only a dream.

3. Cognitive dissonance: a wordless feeling—paralysis, fear, and the visceral sense that there *are* external constraints, and that something is not in accordance with them. If it could be expressed in words, it would be, "I can't jump. It's not possible, it's not right. I'll get hurt." Thus, cognitive dissonance is also a kind of self-protectiveness, self-defensiveness. The dreamed self feels substantial and therefore capable of being hurt. There's a strong identification with the dreamed body, not with the dreamer.

4. Noticing cognitive dissonance: "There's no *reason* for me to feel constrained and thus afraid. There are no external constraints here. I'm only afraid because I'm very used to believing that I can't jump off ledges without getting hurt." In other words, there's awareness that one's negative feeling about the situation isn't produced by the external situation itself, but by the force of one's past experiences with similar situations, a kind of entrenchment in certain patterns of feeling and thinking, or habituation. The fact that one *feels* constraint so strongly in the absence of any actual external constraint is itself noted and seen as surprising and strange.

5. Strengthening conviction about the absence of external constraint: "I know I *can* jump off this ledge without being hurt. This is a dream—the

ground is not solid and hard, and neither is my body, so I can't actually get hurt." The cognitive dissonance is dissipated by habituation to a new set of thoughts about what is real and what is possible.

6. Noticing lack of correlation between will and events (that is, lack of agency): "My body isn't moving the way I want it to, and neither is the landscape. I'm not flying. But it's *my* dream. If I don't control it, who does? Who's in charge here?" There's some discrepancy between the dreamer's will and the dreamed body's actions, and also a discrepancy between the dreamer's will and the dreamed situation (terrain, physical laws, etc.). The self, which is often experienced as one thing, is experienced as many. These selves, though different from one another, are intimately related.

a. First, there is the will, which seems to "belong" to the dreamer. The dreamer's will is the "I" in the sentence, "I notice that I'm dreaming, so I can control my dreamed body's movements in this dream." It's also the "my" in the sentence, "It's *my* dream."

b. Second, there is the dreamed self or dreamed body, which appears to be interacting with things outside of it. When it jumps from a high ledge, the experience of falling and windiness occurs. The movements of the dreamed body correlate tightly with certain sensations. The dreamed self is the "I" in the sentences, "I'm floating to the ground" and "I'm spreading out my arms."

c. Third, there is the uncontrolled aspect of the self, the "givens" of the situation like the terrain, the physical laws themselves, and so on. This aspect seems totally "other" from the dreamed self, but in a lucid dream, the dreamer realizes that these seemingly external and uncontrollable circumstances are not *actually* "other" from the dreamer. The *entire* dreamed situation arises as the dreamer's experience. If the dreamed body is experience, so is the dreamed landscape. The dreamer's will cannot deliberately change certain aspects of the dream, but those aspects are still the dreamer's experiences—they are just experiences that are not controlled by the will. This uncontrolled aspect is the "my" in the sentence, "My mind is creating this whole landscape," and the will is the "I" in the sentence, "There's nothing I can do to change that landscape." The dreamed body is the "I" in the sentence, "I'm walking through this landscape."

7. Noticing the creation of causal mechanisms: "I wasn't wearing a skirt before jumping, but now that I'm falling, I *am* wearing a skirt. My mind must have created this skirt in order to try and account for the fact of my dreamed body's gentle fall." The dreamer had decided that the dreamed self could not be hurt by jumping. Therefore, the dreamer didn't will the dreamed body to

go find a parachute or skirt before jumping, nor did it will a parachute or a skirt into existence. Rather, the skirt is a *causal mechanism* created by the uncontrolled aspect of the mind. It's as if the uncontrolled aspect wondered, "Usually when bodies jump from ledges they fall and get hurt. The dreamed body just jumped from the ledge, and will *not* get hurt. *How* is this going to happen?" The mechanism is something that makes the dreamed body's not getting hurt *fit* with what the mind deeply expects given patterns it knows from waking life (i.e., physical laws). This acts as a kind of defense against cognitive dissonance, a rapid retrospective addition of details to make what has already happened "make sense" according to previously encountered patterns.

If the mind can do all of this while dreaming, perhaps this is more or less what it is doing while *waking*. As strange as dreams might seem, they are not nearly *as* strange as we might expect them to be if we believed that all physical laws and physical constraints were truly inherently *external* to thoughts. That is, in a dream world completely unencumbered by physical laws as experienced in waking life, one still *dreams* physical laws. Dreamed objects and subjects quite often interact according to waking-life expectations. In dreams, in the absence of physical constraints, mental constraints are revealed in all their power. In waking life, we are likely to attribute the tight patterns and thoroughgoing natural order we experience to Nature itself. We tend to believe that mind is supple and matter is recalcitrant. Dreams reveal that even in situations where *matter* is supple, *mind* can be quite recalcitrant.

Any school of thought that remains agnostic about the intrinsic existence of external reality (for example, constructivism, deconstructionism, relativism, skepticism, or contingentism) is often criticized on the grounds that this agnosticism gives far more power to the subject or mind than is due. The specific claim "objects and substances don't exist intrinsically and the parsing of the world into objects and substances is a function of cognition" can be misinterpreted as "there's some void out there" or "stuff disappears when no one's looking at it." Similarly, the specific claim "these objects and substances are not related to one another by intrinsically existing causal laws, but by the cognitive act of noticing that events follow one another regularly" can be misinterpreted as "people can change the world in any way they want to. Physical laws don't apply to people who don't think about them or don't believe in them." The critic's major error is in equating the subject ("person") with the will. A close look at the dynamics of cognition reveals just how difficult it would be to *not* be subject to physical laws!

As the dream example illustrates, very little is controlled by a conscious agent or will. Most of what happens occurs completely unconsciously. The parsing of the world into objects and the relating of those objects into ordered patterns does not primarily happen on the level of intellect or scientific theory—these are only where the finer points of causal stories are worked out. Patterning processes are largely pretheoretical and unconscious. A shift in intellectual orientation and the exercise of conscious will are unlikely by themselves to bring about some radical experiential shift. A contingentist does not attribute as much power to some conscious agent as critics evidently believe or imply. One can say, "I don't believe in the intrinsic existence of objects, so I'm going to stop seeing the world as objects," but it most probably won't work. Developing an intellectual disbelief in substantiality only reaches the most superficial levels of cognition. No one "creates their own world at will." To claim this is to over-identify the subject with the will and to put undue stock in the inherent unity of the subject as a "well-organized organizer," or as the agent that causes things to happen in the world while remaining itself free and *independent*.

In the lucid dream, I mention the awareness of three distinct selves: the will, the dreamed body, and the uncontrolled aspect. Metaphorically, we could describe the self as sailor, ship, and wind, where the will is the sailor, the dreamed body is the ship, and the uncontrolled aspect is the wind. To see the sailor (the conscious will) alone as the self is to carve out only a very small portion of the entire situation. In a dream, the wind, which is totally out of the sailor's control, is still a part of the self—the wind is itself an experience, just like the sailor and the ship are experiences. Similarly, just because something is experienced as not within one's conscious control does not mean that it is totally external to oneself.[5]

ASSUMPTION OF THE INTRINSIC EXISTENCE OF CONTRADICTIONS

The fact that we can experience surprise is often cited as evidence that natural laws must exist, at least to some extent, independently of cognition: "If we're so good at creating order from our experiences, why is that sense of order ever violated? Why are we capable of being surprised by anything? Why do *things continue to happen,* completely independently of whether or not we expected them to happen? And conversely, why do some things *never* happen, no matter how much we expect them to?"

A common response to this question is, "It must be because something independent of our cognition remains outside of our sense of order. This 'something' is the real, autonomous, independent world. It knocks at the door of our cozy, well-ordered theories, and brings us news that we have to *listen* to. This is the basis of new discoveries. Without these surprises, we would just go around and around, stuck in our own closed, self-referential webs of order."

These are, of course, valid anxieties worth addressing. There's something horrifying about the image of ourselves as beings who are impermeable to novelty, as nonlearning creatures, doomed to repetition and the rote following of prior expectation and habit. But the contingentist account does not contend or imply that human beings are that kind of creature.

Notice that the critical response given above pits expectations against *autonomous and radically external reality*. Perhaps a more helpful framing is one that pits expectations against *experience that contradicts expectations*. Here, what counts as a contradiction depends precisely upon expectations. Someone who expects everything would never be surprised. So the sense that an experience is surprising obviously depends on our prior expectations.

Moreover, "contradictory experience" depends not only upon expectations but also upon social practices that define the contours of what is allowed to count as a surprise or, in science, as an anomaly. It actually takes quite a lot to have a contradictory experience. Anyone who has worked in a laboratory knows how many novel and surprising findings are routinely dismissed as conventional and mundane. The term we use for this is often "human error." We have to learn to *discern* when to be surprised and when not. We learn to actually contain our surprise. The novice scientist is enculturated and accepts certain norms of what to be surprised by. This is a critical part of the practice. Much of daily life actually consists of explaining away novel perceptions:

"I was only dreaming."
"They are crazy."
"It was something I ate."
"That squiggle under the microscope is just debris."
"The blot on the picture is noise."
"My eyes haven't adjusted to the light."

A single individual is not entirely *allowed* to be surprised, and certainly not by a single novel perception. When many individuals can communicate with one another about novel perceptions, and can frame those perceptions in a shared language, only *then* is everyone permitted to be surprised.[6]

The world as autonomous reality never forces us, neither as individuals nor as collectives, to adopt certain beliefs or expectations. We convince each other to do so. What scientists offer as factual knowledge does not change in *response* to an autonomous world that somehow *forces* us to respond in any one particular way. People, including communities of scientists, build and shift expectations in response to one another.

These questions of how and when to be surprised are properly *ethical* questions. There is the virtue of "open-mindedness"—the willingness to allow prior expectations to change in response to new experiences (percepts, concepts, etc.). The virtue is tempered by "composure," or the willingness to not be "too" easily convinced. These are things we *value*, and they can be discussed quite coherently in the language of ethics and politics, where we discuss whether these are good values, and why, and how they are to be encouraged or deterred in our common life, and so forth. And actually, these ethical points are mostly delivered as ethical imperatives: "They should be more responsive to evidence" (i.e., they should be *more* willing to change their expectations), and (sometimes in the same breath), "They shouldn't believe everything they see or hear" (i.e., they should be *less* willing to change their prior expectations when faced with new observations).

In sum, that human learning (and organismal learning) happens is not by itself evidence that the universe possesses a particular order to which this learning conforms. It is evidence only that learning happens. We are, thankfully, beings who are capable of being surprised, which is itself an important part of what we call "learning."

ASSUMPTION OF INTRINSIC HIERARCHIES OF ORDER

Phenomena are real, but it is easy to inflate the sense in which they are real. For example, yes, one thing does follow the next. Thus, it is easy to think that something connects them. We do have good predictive technologies, like theories of chemical reactions and certain medical prognoses—so it is easy to think that they are superior to poorly predictive technologies, because they correspond more closely to the intrinsically existent structure of the universe. But good predictive technologies are superior to poor predictive technologies *just because* they help us predict events. This should be good enough, in two senses: first, good enough to satisfy our need to justify our practices, and second, to do all the work, in regard to theories, that we need the word "good" to do. To take it a step further and to say that certain predictive technologies are superior because they correspond more closely to some universal structure is to step into assuming that there *is* an intrinsically existing order.

ASSUMPTION OF A SINGLE ORIGIN AND A LINEAR HISTORY

What about deep time and the origins we narrate as having occurred in the distant past? There is the Big Bang as the origin of physical substance, the first cell as the origin of life, and the first human as the origin of conscious experience (though, depending on how "consciousness" and "experience" are defined, its origin may be placed significantly earlier than human consciousness).

Does a "strange loop" account, one that describes mind and matter as thoroughly interdependent, then say that a preobserver universe did not *exist*? The contingentist claim, predictably, is that the preobserver universe existed, and what this means is that it existed *contingently*.

What is happening when we describe the existence of the preobserver universe? When we say, "The Earth was hot and the air full of smoke 4 billion years ago," we might mean, "If I or another human being had been there, I think it would have seemed like this." Again, just as we cannot say that what *is* is *inherently* a certain way in the absence of an observer who experiences it in a certain way, similarly, we cannot say that whatever *was* was *inherently* a certain way. Cosmological, geological, and paleobiological accounts are important, useful, and grounded in careful and rigorous present-day theory and observations. However, to take these accounts to be saying that the universe was *intrinsically* a certain way at some point in the past raises the same problems as saying that the universe is *intrinsically* a certain way now. To say that sticks "really existed" hundreds of millions of years ago may be to say that they had some inherent substantiality, boundedness, and continuity independent of someone to conceive of them in that way. If this is what is meant by "really exist," then there is no reason to believe in the real existence of sticks hundreds of millions of years ago. If we mean that sticks in the distant past had some dimension, the question arises, "Dimension by what measure and whose measure?" If we mean that sticks had effects on things like water, the question arises, "Who relates to this as a stick and to that as water? Who perceives sticks dropping and simultaneously water rippling, and relates these as 'cause-and-effect'?" History happens—phenomena do give rise to phenomena. Using a dense network of observations and inferences, we might narrate the origin of the Grand Canyon as a stick falling in water a very long time ago: "This happened, then that, then that." These are words and thoughts—very useful ones for certain purposes, and which, if taken too literally, give rise to a reification of past objects and past events. One might get the sense not only that matter exists *outside* of experience exactly as it appears to exist *in* experience, but also that matter existed *before* experience exactly as it appears right now. Such a theory doesn't necessarily bear scrutiny. A seemingly commonsense account that narrates events linearly as "the origin of matter," followed by "the origin of life," followed by "the origin of experience" is not necessarily as sensical as it might seem, and conversely, a strange loop account isn't necessarily as strange as it may seem.

There is an intuition that the origin of "what is" (narrated alternatively as "the origin of the universe," "of matter," "of everything") is a question that can and should be answered, and that the answer requires an explanation of the origin of something from nothing: "Explain why there is something instead of nothing, and when and how that something came into being." Why does this formulation make sense to us? It makes sense insofar as we consider ourselves as the living to be in a privileged position regarding the (intrinsically) real world relative to stones, stars, and the dead. But how could we give ourselves such privilege? By our own physicalist accounts, these inanimate entities both vastly precede and vastly outnumber us, dwarfing our small lives and ideas. It's *them* we want to ask questions of, that world of photons, light years, and event horizons, so that we can access the vastness of time and space that is independent of us, so far beyond our little lives. So we, thickly enrobed in vivid experience, are likely to ask, "Why is there something instead of nothing?" But to the inanimate—that very world we wish to probe and enter and ask the question of—the question is a nonstarter. Ask a star, "Why is there something instead of nothing?" and you're not likely to get an answer. We assume that this is because, for a star, nothing is happening, so there's actually nothing instead of something. So then the question might become, "Explain why something happens for us, and why nothing happens for a stone. Explain when and how that something came into being." This is exactly what the "strange loop" story does: the origin of matter and the origin of experience become the same question. The only way of accounting for the one is to account for the other, to try to narrate in a detailed way how each comes into being dependent upon the other. Why is there something instead of nothing? Because experience arises— absent experience, there *is* no something. Why does experience arise? Because experience comes into being dependent on bodies, dependent on matter, dependent on experience. . . .

The suggestion here is not that cosmological accounts and the problems of finding the origin are not important, interesting, and practical. A great many of us—as theists and atheists, physicalists and dualists, lay physicists and pop psychologists—are engaged in this project, which is clearly meaningful to us. I am suggesting that "something, and its origin from nothing" is not a metaphysically existent entity that demands, by itself, to be explained—any more than a flower is, as Garfield writes, "an entity demanding, on its own, recognition and a philosophical analysis to reveal its essence." Both "something" and "flower" are contingently existent in similar ways.

We are the ones for whom things happen, and we are right to ask, "Why, how, and when do things happen?" However, just because things happen for us, and just because we narrate how things happened in the past ("one thing and then the next, this object and that object"), it does not follow that things happen and things happened in a way that is radically separate from the experiencing or telling of it.

What about memory, which gives the distinct sense that thoughts are being stored somewhere and recalled? In the course of reflecting upon or describing the past ("The other day I went to the store, and . . ."), it is easy to gloss over the fact that this thought of the past is happening *right now*. One can describe right now the experience of what was once "right now."

All memories arise in the present, though they have a referent in the past. A memory is an experience like any other, like a calculation, a perception, or a fantasy.[7] It has a referent (the "contents" of a memory, like the contents of any other experience), just as a name has a referent. However, the referent is not intrinsically existent. Again, the word "flower" stays the same, but no "flower-in-itself" stays the same. Similarly, the referent "past" stays the same, but the "past-in-itself" was not a certain way.

In sum, the idea of the origin of "what is" at a point in time is contingent upon a notion of something in contrast to nothing and does not exist intrinsically. Historical objects exist in the same way, contingent upon notions of contrast, boundedness, continuity, and causal relations. This is not necessarily a matter to be lamented or to feel constrained by, as if the lack of intrinsically existent historical objects means that there is a woeful and intrinsically existent limit to our understanding of history and reality and our ability to comment upon them. This intuition will be examined in more detail later.[8]

ASSUMPTION OF KNOWLEDGE AS LIMITED

An asymptote is a mathematical object, a line that a curve can approach, but never reach (Figure 5.1). The asymptote metaphor of knowledge and reality is that the real world is like a line that our knowledge and experience gets progressively closer to, but never touches. This is a correspondentist view, but it also holds for any pluralist, idealist, or nihilist, who says, "No, we don't think some accounts *correspond* to an intrinsically existent reality more closely than others, but surely there's *something there* (or surely there's *nothing there*) that is independent of us." The something, the nothing, or the irreducibly plural then becomes the line that experience is approaching, but never touches.

We can and do experience our lives as small or limited relative to what is, and this is where the asymptote metaphor gets a lot of its popularity and its punch. It seems that no matter what, we could never fully reach what is. Maybe you yourself feel limited because you've never seen Antarctica, or cannot comprehend the set of real numbers, or are incapable of experiencing sonar as bats and dolphins do, or do not remember being around to witness the Big Bang. Even vast swaths of what is ostensibly *your own life* are utterly opaque to you—everything from gestation through infancy, and the countless uncon-

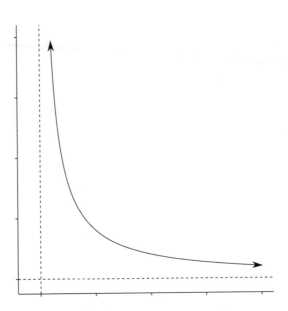

FIGURE 5.1. An asymptote is a line that a curve can approach but never reach.

scious processes that allow you to live at all. To the extent that we think that this is what it means to know what is—to experience the sum of all theoretically possible experiences—of course, we will always fall short. This "theoretically possible," this *imagined* space, then keeps us separate from what *is*. We are then likely to see ourselves as beings with inherently limited knowledge, in a world populated by the same: infants who experience bottomless abandonment when they're left alone in a room, rodents who eat their young when their scent is changed, and baffled goldfish swimming around the same little bowl.

But there may be a different way of understanding what it means to "know what is." Yes, a world arises dependent upon processes like embodiment, sensing, and cognition—*and* it is certainly not necessary that these processes be experienced as *limits*. These processes do not, after all, produce *an inherently limited world*. They produce *a world*, full stop. To call that world inherently limited is again to assume that there is an intrinsically existent reality compared with which our worlds—and indeed, the worlds of every living being—are small. The idea of a cognitive limit is just that: an idea. We think the cognitive limit is that our experiences do not have the full content of all possible experiences, analogous to our lives as being a mere subset of the vast space of all that is theoretically possible. But maybe the cognitive limit is that we assume there's more to it all than just this: more to objects than vividness and predictability, more to subjects than experiences, more to the natural order than conceptual patterns, and more to knowledge than the ongoing process of living itself.

Considering the myriad things upon which our knowledge depends, we may feel trapped, doomed: "Help! I can't get out of my body, my senses, my thoughts, my language, my culture, my society, my environment!" But we can ask ourselves, was it supposed to be some other way? Aren't these exactly what allow anything at all to happen for us? Don't our experiences depend completely upon these? Another possible response is, "Thank goodness! Here are my body, my senses, my thoughts, my language, my culture, my society, my environment! And therefore, here I am, here it all is!" In other words, instead of experiencing these processes as *constraints* upon some abstract and imaginary theoretical space—a picture that, despite good intent and limited usefulness, becomes easily reified and used to torture ourselves and others into both lifelong *separation from* and *desperation for* the really, really real world—we can experience these processes as precisely what *allow us to live* and to live in a world that is, yes, plenty vast, plenty vivid, and changing endlessly as the processes themselves change.

In sum, the gap that is assumed to exist between organismal perception and reality (the world as it is) *is* a gap between subjects and objects—but subjects, objects, and this gap between them arise *contingently*, and they do not exist inherently. There is no intrinsically existing gap between subjects and objects, or between experiences and worlds. The gap arises contingent upon the experience of there being such a gap. This entire book has so far been an account of some of the many pre-theoretical assumptions (called "assumptions," "inferences," or "intuitions" throughout) that can give rise to the experience of this gap.

Again, the question, "How do organisms sense the world?" is often read by physicalists as, "How do organisms, with various degrees of accuracy or completeness, sense the one real world that exists independent of all of their perceptions?" Contingentists maintain that the real world is available to organisms but abandon the notion of *intrinsic* reality. The world is not an illusion. It is precisely what it is, which is nondifferent from experiences. Organisms do indeed experience the world as it is, *and* those experiences are various. Experiences are various because bodies are various. Similar bodies, shared behaviors, comparable modes of patterning, and coordinated actions give rise to common experiences. In this sense, a contingentist reads the question "How do different organisms sense the world?" as "What *is* the world?" To say very much the same thing, by describing the diversity of bodies, behaviors, and perceptions, contingentists see biologists as expanding our understanding of what the world is—that is, vast, varied, and inexhaustible.

———

What does natural order depend on? It depends upon practices of patterning themselves. It depends on cognitive consonance, or the resolution of unexpected experiences into experiences that "make sense." It depends on the capacity to know when and when not to be surprised in the face of novel experiences. It depends on the valuing of theories based on virtues like predictive power, but not necessarily based on their assumed correspondence to an intrinsically existent order. Natural laws and natural objects did not necessarily originate as intrinsically existent quantities or states of affairs at a particular point a very long time ago. Such an account assumes the intrinsic existence of *something* in contrast to *nothing*, but both of these exist contingently. In foregoing the notion of natural order *as an intrinsically existent state of affairs,* the contingentist view does not refuse reality. Rather, contingentists preserve the world's reality *just by* their refusal to posit an order that is radically external to subjects, a truth that perceivers will never attain, or a reality from which subjects are forever separated.

CONCLUSION

Life as We Know It

This book thus far has been a reckoning with many absences:

> the absence of referents independent of terms,
> the absence of objects independent of perceptions,
> the absence of essences within things,
> the absence of causal powers between regularities,
> the absence of subjects independent of experiences and actions,
> the absence of laws independent of concepts and cognitive consonances, and
> the absence of gaps between subjects and reality independent of the experience of such.

There is also the absence of absence itself, insofar as "absence" is itself a *term*, which brings us full circle back to the absence of referents independent of terms. That is, there is no intrinsically existent referent to the term "absence"— no intrinsically existent Nothing or Void. This is how contingentism differs from nihilism, and this is why contingent existence can be considered genuine, bona fide existence, and not a kind of "lesser" existence. These "absences" of which I speak are thus, of course, just the absence of *inherent* existence.

THOROUGHGOING CONTINGENCY AND THE ABSENCE OF INHERENT EXISTENCE

The absence of inherent existence, then, is simply another way of illustrating the thoroughly *contingent* existence of things, as explained in detail in Chapter 1. In closely examining the object and the subject of the process called "sensing" (or perceiving, or experiencing—or, simply, *living*), we find no inherently existing object, no inherently existing subject, and no inherently existing process. Instead, all of these designations are inseparable from one another—they cannot exist on their own, by their own "power," inherently. Rather, they bring each other about:

1. Objects exist as objects dependent on subjects and processes. Subjects through perceptual-conceptual-behavioral processes (sensing, feeling, thinking, acting) unify phenomena and differentiate them from other phenomena, thus giving rise to objects.

2. Subjects exist as subjects dependent on processes and objects. Subjects unify *themselves* through perceptual-conceptual-behavioral processes and differentiate themselves from other phenomena, called objects. The subject as *agent* perceives itself as having causal powers. When we say "the subject perceives itself," there is no intrinsically unified object that is perceived, and no intrinsically unified subject doing the perceiving.

3. Perceptual-conceptual-behavioral processes exist dependent on subjects and objects. Subjects narrate these processes as links between themselves and objects from which they are separate. A complex network of terms and concepts ("object," "energy," "information," "form," "transformation," "assimilation", "transduction") sustains both the sense of *separateness* of objects from subjects and the sense of *interaction* of objects with subjects.

In short, not the object or the subject or the process that links them is firm ground. None of them exists intrinsically, so none of them can be the fundamental support for the others. Their stability arises precisely from their interdependence—they keep each other in place contingently. The idea that experiences arise without intrinsically existing experiencers or some intrinsically existing process of experiencing can appear strange at first (hence the "strange loop" described earlier), but is not an illogical or inconsistent position. On one hand, the idea of "experiences without ground" can give a sense that what *is* is very spacious, open, and free. It conveys the sense that *nothing* exists intrinsically: neither matter, nor mind, nor time, nor space, and therefore, everything is mutable and a great deal is possible. On the other hand, the idea of "experiences arising from tight, self-referential interdependence" gives a sense that what *is* is narrow, closed, and constrained—everything holds everything else in place, such that all arises in a regular way, with alternative happenings being difficult or impossible. These are two sides of the same coin.

Like all scientists, biologists seek patterns and describe them. Patterns relate to other patterns, and biology becomes an increasingly intricate, beautiful, regular, and dense net of relationships. It is easy to imagine that the net is grounded somewhere. Sometimes when we come across a spider's web, it can be difficult to find where it's anchored; yet the assumption is that it is anchored somewhere. Similarly, it is easy to assume that the dense net of experiences is anchored somewhere—in a world of objects, or in a body, brain, or soul. We often believe that the regularities we experience must be grounded in some kind of substance *beyond* them—material, spiritual, or mental. However, it is entirely possible that the net is aloft, that it is not tethered to anything outside of it. In fact, as far as anyone can tell, the net is *all there is*, so there can be nothing outside of it that could serve as a tether.

GENEALOGIES, RELATIONS, AND INTELLECTUAL KINDRED

Having illustrated with many examples throughout this book how contingentism "works," I return to a question that may have seemed salient at the beginning of the book: Why "contingentism," when other terms have already done a great deal of work detailing the interdependence of objects and subjects? After all:

Doubtfulness about the status of truth claims as correspondence to intrinsically existent categories is already "skepticism" (Sextus Empiricus, Pyrrho).

Justification of claims and practices with reference to their *usefulness* (instead of with reference to correspondence to any intrinsically existent category) is already "pragmatism" (William James, Richard Rorty).

The interpenetration of parts and wholes is already "dialectics" (Karl Marx, Richard Lewontin, Richard Levins).

The specific tracing of the coming-into-being of things at the nexus of myriad occurrences is already "genealogy" (Friedrich Nietzsche, Michel Foucault).

The co-construction of object and subject is already "constructivism" (Ludwik Fleck, Barbara Herrnstein Smith).

The process by which objects and subjects come into being at all is already "ontogenesis" (Gilbert Simondon).

The "bringing forth of a world by living" is already "autopoeisis" (Humberto Maturana, Francisco Varela, Evan Thompson).

The development of organisms as contingent and inseparable into intrinsically existent categories like "nature" and "nurture" is already "developmental systems theory" (Susan Oyama).

The refusal to reify the difference between matter and mind is already "nondualism" or "dependent arising" (Nagarjuna, Jay Garfield).

The analysis of metaphysical entities to reveal what they themselves depend on is already "deconstruction" and "rhizomatics" (Jacques Derrida, Gilles Deleuze).

The absence of an intrinsically existent fundament upon which all else depends is already "irreductionism" (Bruno Latour).

What, then, is the use of "contingentism" in this already large family of terms?

First, "contingentism" is one way of bringing the family together by emphasizing an important common thread between them (a family resemblance, perhaps). These positions are each distinct with respect to their history, usage, and claims, and none of them overlaps exactly with the claims of contingentism (summarized in Chapter 1, "Features of contingentism"). However, they do share a commitment to emphasizing the mutual dependence of ontological categories, thereby not taking for granted their *independence*.

Second, by drawing attention to this common thread of mutual dependence, we can see its relative simplicity. To put it very simply: Yes, everything *does* depend. And by everything, we mean *everything*, without remainder. The elegance of this formulation may render this important component of each of the aforementioned positions quickly intelligible.

Third, and finally, "contingentism" also suggests natural alliances with the biological sciences. There are already several concepts of contingency-as-dependence in circulation in biology:

1. Soil and sun produce grass, which is eaten by deer, who are eaten by wolves. All of these are decomposed by microbes, which make nutrients available to grass. Living beings depend on each other and on nonliving factors in order to continue living. This is an example of the interdependence of living beings, and of living beings and nonliving entities, or *ecological interdependence*.

2. An enzyme is produced dependent upon the presence of the molecule that it cleaves. The enzyme lactase, for example, is produced in the presence of the sugar lactose, which it digests. When the lactase has cleaved all of the lactose, it is no longer produced, thereby effectively "shutting off" the cycle. This is an example of the interdependence of products and processes, or *regulatory interdependence*.

3. A cell population metabolizes and excretes chemicals in a way that acidifies the local environment, producing changes in the composition of the cells themselves. A smaller-scale entity influences a larger-scale entity, which in turn influences the smaller-scale entity. This is an example of the interdependence of parts and wholes, or *hierarchical interdependence*. This kind of interdependence is also at play in human and animal physiology, and in the medical sciences—for example, the heart circulates blood throughout the entire body, which in turn provides oxygen and nutrients to the heart.

To this (nonexhaustive) set, the use of contingency employed in this book adds and explores a fourth type of interdependence, one which has historically not been emphasized in most biological accounts. This is mutual constitution, or *ontological interdependence*—the view that things only exist *as* things dependent upon other things.

Many biologists are ecological thinkers, acknowledging the constant interaction of organisms and environments, and their close interdependence. Standard interactionism, however, divides the world into objects and subjects, and then says, "They're always interacting." This division need not necessarily be taken as an inherently existing prior. Contingentism in the sense of ontological interdependence is a more thoroughgoing ecological worldview, one that acknowledges the *total* interdependence of phenomena, and thus the absence of the inherent existence of either subject or object, either organism or environment. In this sense, contingentism simply follows down the full implications of a way of thinking that is already familiar to many biologists.

THE MANY FORMS THAT WONDER TAKES

You may have heard the phrase, "It's all connected" before. What does it even *mean?* In this book, we have explored the view that interdependence can mean more than viewing things as *connected*. Rather, as we have discussed in detail, things can be viewed as dependent on one another in such a way that *each brings the other into being.*

Taking interdependence seriously means not simply accepting it as a truism. Contingentism is at its most useful, then, not so much as a set of claims or position statements but as a practice, a habit, and a method: the constant asking of the question, "What does this depend on?" This can *feel* like a practice—that is, an action or an operation, something that requires involvement and effort—because it goes counter to the usual, familiar, and therefore seemingly effortless practices in which many of us are engaged most of the time. We have discussed these "effortless practices," or assumptions, at length throughout the book, and through the examination have come to a perhaps counterintuitive but, I hope, useful and satisfying perspective.

From this perspective, phenomena can seem like mirages—vivid in appearance, but apt to disappear when we examine them closely. In one sense, the word "mirage" is misleading because our intuition is that a mirage is "false." Yet the word "mirage" derives from the Latin *mirus,* meaning "wonderful," "amazing," or "astonishing." The word does not contrast with the true. It simply has the connotation of "that which inspires wonder." Mirages are mirages—they are not intrinsically water, but they arise as water dependent upon eyes, observers, distance, and bright light. Similarly, phenomena are phenomena—they are not intrinsically objects or subjects, but they arise as objects and subjects dependent upon many conditions. And this very happening is indeed quite wonderful, amazing, and astonishing in its way.

In studying life and living beings, as biologists both professional and lay, we may find ourselves struck by many different kinds of wonder:

> In examining the interlockings of many parts arranged in a functioning whole, from mechanical clocks to computers to dividing cells, we might think, "How *neat*—look at how beautifully they function and work all together."
>
> In noticing the great multiplicity of things, like the many types of molecules and beaks and beings, we might think, "How *awesome*—what an astonishing feast of variety, diversity, and complexity here."
>
> In tracing the many happenings upon which our lives depend, from photosynthesis to the gentle movements of our stomachs, we might think, "How *touching*—what would any of us do or be without all this and without one another?"
>
> In formulating or comprehending a general principle, being *struck* by it for the first time, we might think, "How *elegant*—there's such simplicity and clarity in this explanation (or model, or equation), and I feel that the whole world is right here, at hand, in this small phrase."

In addition to these modes of wonder, the contingentist view suggests a few more:

> "How *new*—not one thing changing, not many things transforming one another, but all altogether new: now, and again now."
>
> "How *strange*—it all loops back on itself. I walk on land that's outside of me only to find that it's not strictly outside of me, but is instead critically dependent on me. I search my insides for the place where the land comes from, but there's no place and no source to be found there either. The whole thing is aloft. And yet: here it is."

CODA: SMALL, VAST WORLDS

There are two common ways of compartmentalizing mystery so that it doesn't enter our comfortable, predictable, everyday world. One is to put it in the realm of "inexplicable things that no one will ever know," as dualists tend to do. Another is to put it in the realm of "currently inexplicable things that we may know some day in the future," as physicalist monists tend to do. Both approaches take the mystery out of mundane existence, the latter by eliminating it altogether, the former by relegating it to a separate, sacred sphere.

But we do not need to separate the material from the mystical, nor do we need to eliminate either the material or the mystical from existence. The material world and the mystical world could be *exactly* the same place in every respect. For one, the world of cold, unfeeling, and inanimate matter is already vibrantly *animated* by our instinct that matter is, by itself, capable of being cold, unfeeling, and inanimate. Furthermore, a predictable world is still absurd, an insubstantial world is still vivid, and a shared world is still multifarious. A world of experiences is—*genuinely*—a world of real things. A world of real things is—*truly*—a world of experiences. And finally, a world that is vast—so vast that it is simply not a *thing* that can be sensed, remembered, predicted, comprehended, or even imagined; a world that seems like it should therefore, by definition, be impossible for us as the limited living to reach—that very world is just this, where you and I and the eight-celled algae live every moment, right here.

NOTES

1. IT DEPENDS: EXISTENCE AS CONTINGENT

1. A. Batko and J. Jakubiec, "*Gonium dispersum*, a new species of *Gonium* from Poland," *Archiv für Hydrobiologie Supplement* 82, no. 1 (1989): 39–48.

2. "Multicellular" in organismal and evolutionary biology has several specific technical meanings. Some biologists contend that in order for an organism to be multicellular, (a) the constituent cells cannot live separately from the rest of the organism and (b) germ cells must be differentiated completely from soma (body) cells so that only germ cells can give rise to a new organism. (In humans, for example, only sperm and egg cells can develop into a baby; neurons and skin cells cannot). For the purposes of this discussion, I use "multicellular" less strictly to mean an organism composed of many, differentiated, mutually dependent cells. For a comprehensive and accessible review of some of the issues surrounding transitions to multicellularity, see David L. Kirk, "A twelve-step program for evolving multicellularity and a division of labor," *BioEssays* 27, no. 3 (2005): 299–310.

3. For those with two particular color receptors in their eyes (such as most mammals and colorblind people), red and green hues are indistinguishable. Most people have three color receptors and can distinguish between red and green hues.

4. Humans, unable see light in the ultraviolet range, see daisies as yellow centers with petals of an undifferentiated white color. Bees, which can see in the ultraviolet range, can discern bulls-eye patterns on the petals which are invisible to humans, but do not necessarily distinguish the colors of the daisy's center and its petals. Bjørn Rørslett (NN/Nærfoto) takes photos of flowers under ultraviolet light, revealing patterns not usually seen by humans. See http://www.naturfotograf.com/index2.html.

5. J. Garfield, *The Fundamental Wisdom of the Middle Way: Nagarjuna's Mulamadhyamakakarika* (Oxford: Oxford University Press, 1995).

6. The word "label," as I use it here, or "attribute," which I use elsewhere, are too strong, in that they convey the sense of some willful action on the part of an agent—"the observer labels." I do not wish to reify "the observer" in any way, granting it in-

herent existence and / or the causal power to make objects through "mere" labeling. I make this quite clear in Chapter 4, where I focus on the contingent existence of the observer itself. One of the difficulties of writing a piece like this is that in order to highlight the dependence of one phenomenon on another (in this case, the dependence of an object like a flower on the cognitive activities of the observer), I need to temporarily quasi-reify or "hold still" as an entity whatever is depended upon (in this case, the observer itself).

7. The use of "seamless" here does not connote a kind of monism. "Seamless" in this context means "not intrinsically different or disjunct," yet it does not mean "one thing changing."

8. R. Rorty, *Philosophy and the Mirror of Nature* (Princeton: Princeton University Press, 1979).

9. Monism(s) and dualism and their relationship to contingentism are discussed in greater depth in Chapter 4.

10. The "law of attraction" was popularized in Rhonda Byrne's self-help book *The Secret* (New York: Atria Books, 2006), which holds that one's thoughts are capable of "attracting" one's physical circumstances. Positive thinking, by this formula, can attract positive outcomes, not simply by increasing psychological well-being but also by physically attracting new objects and states into one's life. Thinking about getting a new house and acting in ways as if it were already there can thereby bring the house into one's life. Again, contingentism is not about these kinds of changes in matter based on changes in mind.

11. For reviews of current issues in cognitive dissonance and social psychology, see J. Cooper, *Cognitive Dissonance: 50 Years of a Classic Theory* (London: Sage, 2007); and B. Gawronski and F. Strack, eds., *Cognitive Consistency: A Fundamental Principle in Social Cognition* (New York: Guilford Press, 2012).

2. WHAT DO OBJECTS DEPEND ON? PHYSICAL SUBSTANCE, MATTER, AND THE EXTERNAL WORLD

1. In using "humans," I do not mean to negate crucial cross-cultural and inter-individual differences that exist in these processes, nor am I arguing for a species-universal human folk physics or folk materialism. Indeed, as Bruno Latour has argued in both his *We Have Never Been Modern* (Cambridge, MA: Harvard University Press, 1993) and his more recent *An Inquiry into Modes of Existence: An Anthropology of the Moderns* (Cambridge, MA: Harvard University Press, 2013), assumptions about an intrinsically existent external "nature" are made characteristically by those who consider themselves "modern." I remain agnostic regarding the question of exactly *who* and *how many* hold these assumptions, and I direct the reader to Latour's works for his views on this important question. Nonetheless, I concur with Latour that at the very least, "moderns" tend to hold these assumptions, and I imagine many of the readers of this book to be moderns. This is why I say *"common* inferences that humans make" (not universal), and "inferences that humans *tend* to make" (not all humans, and not always).

There is a vast literature in anthropology and cognitive sciences regarding cross-cultural and inter-individual modes of discerning objects, to which the reader is directed for more on this intriguing subject. See G. E. R. Lloyd's *Cognitive Variations* (Oxford: Oxford University Press, 2009) for a helpful review of the claims, findings, and limits of cross-cultural studies in cognitive science.

2. As a crude example of how objects are not "self-presented" or "obvious to anyone," consider that one can conceivably argue for or against a stem being a part of a flower.

3. The most familiar exception to this is nuclear fusion in the core of stars.

4. Atomic dimension is usually expressed as the size of the probability cloud of the electrons of the atom. In other words, the atom's outer limit is the volume around the nucleus where the outermost electron is predicted to be found with just under 100 percent confidence. As outermost electrons are gained and lost in the common chemical interactions between atoms, the sizes of "atoms themselves" decrease. In the discussion of particle size, for the sake of simplicity, I consider atoms without electrons (fully ionized atoms, i.e. just atomic nuclei).

5. J. Garfield, *The Fundamental Wisdom of the Middle Way: Nagarjuna's Mulamadhyamakakarika* (Oxford: Oxford University Press, 1995).

6. For an introduction to quantum electrodynamic theory, see Richard Feynman, *QED: A Strange Theory of Light and Matter* (Princeton: Princeton University Press, 1989).

7. For his criteria of causes, see David Hume, *A Treatise of Human Nature* (New York: Macmillan and Co., 1888).

8. Ibid.

9. See Garfield (1990) for an important and careful argument against causes as "glue" (i.e., the third thing that links two events). Garfield illustrates the prevalence and problems of this "glue" view of causation in the sciences and argues that causation is just one of the realms where the sciences may benefit from insights from traditions of philosophical skepticism.

10. Critiques of a regularity view of causation are not being criticized here. Indeed, probabilistic views of causation, such as those being used in quantum thermodynamics, can be very useful.

11. See Coventry (2006) for a particularly careful and interesting reading of Hume as a "quasi-realist."

12. H. R. Maturana and F. Varela. (1992). *The Tree of Knowledge: The Biological Roots of Human Understanding*. Boston: Shambhala Publications.

13. There is surely a great deal of variability between individuals and between collectives with respect to this kind of experience. An experience that is not open to collective observation (a personal dream, vision, or so forth) can still be experienced as intersubjectively real, particularly if one's social group considers such experiences *as* part of the "really real world." Indeed, as Latour argues (particularly in *An Inquiry into Modes of Existence*), the metaphysical view that dreams, visions, and the like are *separate from* or *do not belong* as part of the "really real world" is a peculiar habit of moderns.

14. For a useful elaboration of this idea through a constructivist reading of the work

of Ludwik Fleck, see Barbara Herrnstein Smith, *Scandalous Knowledge: Science, Truth and the Human* (Durham, NC: Duke University Press, 2006), 46–84.

15. For a similar argument regarding the metaphor of a "hostile" nature in scientific writing, see Evelyn Fox Keller, *Secrets of Life, Secrets of Death* (New York: Routledge, 1992), 116–119.

16. Of course, concepts are defensible, just not on the grounds of correspondence to an intrinsically existent state of affairs.

17. Contingentism, of course, tends to be charitable toward belief, not thinking of belief as "mere" at all (in comparison, say, to genuine knowledge). The point here is simply that if one wishes to preserve the idea of beliefs as mere, one can still consider the world's necessary independence from mere belief to be a mere belief itself.

18. I steer clear of quantum physics and the apparent "spookiness" of the famous double slit experiment because I would be out of my depth in commenting on these. More important, I have meant to show the relevance of contingentism in the *absence* of accounts from quantum physics. I appreciate the importance of these experiments and theories; however, the interdependence of objects and subjects that I mean to illustrate does not depend on things like the wave function of an electron being collapsed during its observation. Again, the emphasis here is on folk physics, or the appearance and interpretation of objects in everyday life.

3. WHAT DOES SENSING DEPEND ON? TRANSDUCTION, ENERGY, AND THE MEETING OF WORLDS

1. K. J. Hellingwerf et al., "Current topics in signal transduction in bacteria," *Antonie van Leeuwenhoek* 74 (1998): 211–227.

2. D. Burbulys, K. A. Trach, and J. A. Hoch, "Initiation of sporulation in *Bacillus subtilis* is controlled by a multicomponent phosphorelay," *Cell* 64 (1991): 545–552.

3. B. L. Wanner, "Is cross regulation by phosphorylation of two-component response regulator proteins important in bacteria?" *Journal of Bacteriology* 174 (1994), 2053–2058.

4. S. J. H. Ashcroft, J. R. Crossley, and P. C. C. Crossley, "Effect of *N*-acylglucosamines on the biosynthesis and secretion of insulin in the rat," *Biochemistry Journal* 154 (1974): 701–707.

5. D. G. Pipeleers, M. Marichal, and W. J. Malaisse, "The stimulus-secretion coupling of glucose-induced insulin release. xiv. Glucose regulation of insular biosynthetic activity," *Endocrinology* 93 (1973): 1001–1017.

6. A.-H. Maehle, C.-R. Prull, and R. F. Halliwell, "The emergence of the drug receptor theory," *Nature Reviews: Drug Discovery* 1 (2002): 637–641.

7. J. N. Langley, "On the reaction of cells and of nerve-endings to certain poisons, chiefly as regards the reaction of striated muscle to nicotine and to curare," *Journal of Physiology* 33 (1905): 374–413.

8. A.-H. Maehle, C.-R. Prull, and R. F. Halliwell, "The emergence of the drug receptor theory," *Nature Reviews: Drug Discovery* 1 (2002): 637–641.

9. O. Hammarsten and J. A. Mandel, *A Text-book of Physiological Chemistry*, 4th ed. (New York: John Wiley & Sons, 1904).

10. Ibid.

11. Ibid.

12. A. J. Clark, *The Mode of Action of Drugs on Cells* (London: Edward Arnold & Co., 1933).

13. R. P. Ahlquist, "A study of the adrenotrophic receptors," *American Journal of Physiology* 155 (1948): 586–600.

14. E. Hildebrand, "What does Halobacterium tell us about photoreception?" *Biophysical Structural Mechanics* 3 (1977): 69–77.

15. B. D. Gomperts, I. M. Kramer, and P. E. Tatham, *Signal Transduction* (Orlando, FL: Academic Press, 2002).

16. M. Rodbell, "Signal transduction: Evolution of an idea." Nobel lecture, 1994. Retrieved on June 5, 2009, from http://nobelprize.org/nobel_prizes/medicine/laureates/1994/rodbell-lecture.pdf.

17. Ibid.

18. Ibid.

19. "Reactivity" and "behavior" are other terms used in addition to "responsiveness," though not fully interchangeably.

20. Hammarsten and Mandel, *Text-book of Physiological Chemistry*.

21. J. Adler and W-W. Tso, "'Decision-making' in bacteria: Chemotactic response of *Escherichia coli* to conflicting stimuli," *Science* 184, no. 4143 (1974): 1292–1294.

22. N. Wiener, *Cybernetics: Or Control and Communication in the Animal and the Machine* (Cambridge, MA: MIT Press, 1948).

23. L. Fleck, *Genesis and Development of a Scientific Fact* (Chicago: University of Chicago, 1979).

24. Indeed, noninertial movement (where inertial movement means movement where objects will move in a straight line unless stopped) is one of the key features of agency attribution. In a landmark experiment in 1944, psychologists showed subjects an animation of geometric figures exhibiting noninertial movement and found that viewers very quickly and easily attributed behavior, agency, even personality to simple shapes. See F. Heider and M. Simmel, "An experimental study of apparent behavior," *American Journal of Psychology* 57 (1944): 243–249.

25. The rapid influx and efflux of ions changes the charge of the cell and causes a signal—an action potential—to be passed on to another cell. This is essentially a kind of "electrical signal," in that it relies on a transfer of charges.

26. R. H. Douglas et al., "Dragonfish see using chlorophyll," *Nature* 393 (June 4, 1998): 423–424.

27. I. Washington et al., "Chlorophyll derivatives as visual pigments for super vision in the red," *Photochemistry and Photobiology Sciences* 6, no. 7 (2007): 775–779.

28. I do not wish to endorse here the view that light precisely *initiates* a visual event. Vision depends upon light, but it depends as much upon what is already happening with (not *in*) the brain and body. For a cogent analysis of vision as a function of whole

brain and body dynamics, see F. J. Varela, E. Thompson, and E. Rosch, *The Embodied Mind: Cognitive Science and Human Experience* (Cambridge, MA: MIT Press, 1991).

29. R. H. Douglas, C. W. Mullineaux, & J. C. Partridge. (2000). Long-wave sensitivity in deep-sea stomiid dragonfish with far-red bioluminescence: evidence for a dietary origin of the chlorophyll-derived retinal photosensitizer of Malacosteus niger. *Philosophical Transactions of the Royal Society B: Biological Sciences, 355*(1401), 1269–1272.

30. Thankfully, there is a fascinating and growing literature offering alternatives to the division of organisms into their "merely" bodily and "interestingly sophisticated" cognitive functions. See H. R. Maturana and F. Varela, *The Tree of Knowledge: The Biological Roots of Human Understanding* (Boston: Shambhala Publications, 1992); A. Noë, *Out of Our Heads: Why You Are Not Your Brain and Other Lessons from the Biology of Consciousness* (New York: Hill and Wang, 2009); and L. Barrett, *Beyond the Brain: How Body and Environment Shape Animal and Human Minds* (Princeton: Princeton University Press, 2012).

31. J. Jeffery, Lecture notes, "Introduction to energy," http://www.nhn.ou.edu / ~jeffery / course / c_energy / energy1 / lec001.html. Retrieved November 29, 2009.

32. R. P. Feynman, *The Feynman Lectures on Physics, Volume 1* (Redwood City, CA: Addison-Wesley, 1964).

33. Compare this refusal to reduce phenomena to a single intrinsically measure to "irreductionism," described in Bruno Latour's *The Pasteurization of France* (Cambridge, MA: Harvard University Press, 1993).

34. See Garfield (2015) for an important discussion of the history of this particular view of agency, and its implications for the development of contemporary Western ethics and law.

35. See biologist Jacques Monod's classic work on just the topic of agency emerging from a determined universe: *Chance and Necessity: An Essay on the Natural Philosophy of Modern Biology* (New York: Knopf, 1971).

36. Panpsychism is an umbrella term for a diverse school of thought. The Stanford Encyclopedia of Philosophy (SEP) defines it, helpfully, as "the doctrine that mind is a fundamental feature of the world which exists throughout the universe." Some thinkers take this to mean that all entities in the universe have mental properties; others take it to mean that the universe is itself a mind and its parts may or may not have mental properties. The SEP entry is an excellent introduction to the field: W. Seager and S. Allen-Hermanson, "Panpsychism," *The Stanford Encyclopedia of Philosophy* (Fall 2013 Edition), ed. E. N. Zalta, http://plato.stanford.edu/archives/fall2013/entries/panpsychism. See also Charles Birch's *Biology and the Riddle of Life* (Sydney: University of New South Wales Press, 1999) for a linking of panpsychism with diverse questions in biology and theology, and A. N. Whitehead's classic work addressing panpsychism, *Process and Reality: An Essay in Cosmology* (New York: Macmillan, 1929).

37. This view of reciprocal co-construction is best exemplified in niche construction theory, a view in biology that organisms construct their environments, which in turn change the development and evolution of organisms. See F. J. Odling-Smee,

K. N. Laland, and M. F. Feldman, *Niche Construction: The Neglected Process in Evolution* (Princeton: Princeton University Press, 2003).

4. WHAT DO ORGANISMS DEPEND ON? BODIES, SELVES, AND INTERNAL WORLDS

1. M. J. West-Eberhard, *Developmental Plasticity and Evolution* (New York: Oxford University Press, 2003).

2. Cited in R. Ottolengui, ed., *Items of Interest: A Monthly Magazine of Dental Art, Science and Literature* 33 (1911).

3. The question of *which* organisms do this is beyond the scope of this work. See S. Savage-Rumbaugh, S. G. Shanker, and T. J. Taylor, *Apes, Language, and the Human Mind* (New York: Oxford University Press, 1998), F. de Waal, *Primates and Philosophers: How Morality Evolved* (Princeton: Princeton University Press, 2006), and L. Barrett, *Beyond the Brain: How Body and Environment Shape Animal and Human Minds* (Princeton: Princeton University Press, 2012) for more regarding nonhuman cognition.

4. A. Cleeremans, ed., *The Unity of Consciousness: Binding, Integration, and Dissociation* (New York: Oxford University Press, 2003).

5. See A. Brook and P. Raymont, "The unity of consciousness," *The Stanford Encyclopedia of Philosophy* (Spring 2014 Edition), ed. Edward N. Zalta, http://plato.stanford.edu/archives/spr2014/entries/consciousness-unity.

6. M. Botvinick and J. Cohen, "Rubber hands 'feel' touch that eyes see," *Nature* 391, no. 6669 (February 1998): 756.

7. This includes "talking to oneself"—the "I" can indeed have detailed social interactions with "other" aspects of the "I"!

8. R. Rorty, *Philosophy and the Mirror of Nature* (Princeton: Princeton University Press, 1979).

9. J. Butler, *Precarious Life: The Powers of Mourning and Violence* (London: Verso, 2006).

10. See A. Noë, *Out of Our Heads: Why You Are Not Your Brain, and Other Lessons from the Biology of Consciousness* (Hill and Wang: New York, 2009), for a cogent analysis of the problems with the search for self, world, and experience *within* the brain.

11. I should be careful to define what "necessary" means here. Again, there are no causal powers, just regularities. Physical substance does not have some inherent *causal power* to create organisms. The regularities here are that no organisms are observed in the absence of physical substance, no physical substance is observed in the absence of thought, and so on. In other words, again, when this happens, that happens, but this does not somehow *make* that happen.

12. D. Hofstadter, *Gödel, Escher, Bach: An Eternal Golden Braid* (New York: Basic Books, 1979), and I *Am a Strange Loop* (New York: Basic Books, 2007). The "strange loop" that Hofstadter references in *I Am a Strange Loop* is self-reflexivity, which creates the self. His analogy is that identity is like the image that appears on a TV screen that displays the output of a video camera pointing to that very TV screen. The video camera is self-consciousness, and the TV screen is the brain. He has no problem with the

fact that the "screen" (the brain) is itself made of neurons, themselves physical substance. The loop I evoke, in contrast, is thoughts about the existence of their own physical support. In other words, a brain is itself a thought and the thought attributes its own existence to a brain ("I think I'm thinking, I think I have a brain, and I think this thought is coming from that brain"). Hofstadter seems to be a physicalist monist, saying that ultimately or fundamentally it is *matter,* properly arranged, which gives rise to (or *is*) consciousness—not the other way around, and not fully mutually constitutive. He does acknowledge the puzzling aspects of this formulation.

13. This theme of *recursion* appears in a number of important theories of life and cognition, such as autopoiesis, second-order cybernetics, and systems theories in general. An important subset of these fields places its emphasis primarily on *organisms* or *bodies* as themselves recursive, self-maintaining (or autopoietic) systems, where organisms are themselves a closed network of mutually "enabling conditions." See E. Di Paolo and E. Thompson, "The enactive approach," in *The Routledge Book of Embodied Cognition,* ed. Lawrence Shapiro (New York: Routledge, 2014). The use of the word "closed" here or "closure" does not refer independence from energy or material resource flows, but is rather a reference to the topology or *organization* of living organisms; see B. Clarke and M. Hansen, "Neocybernetic emergence: Returning the post-human," *Cybernetics and Human Knowing* 16, nos. 1–2 (2009): 83–99, for an important and clarifying discussion on the specific use of "closure" in second-order cybernetics and systems theories. A famous illustration of this "principle of biological organization" appears in Humberto Maturana and Francisco Varela's *The Tree of Knowledge* (Boston: Shambhala Publications, 1992), where the authors note that the cell membrane is both a product of cellular metabolism (in that the membrane must be created and maintained by biosynthesis of lipids and proteins) and a cause of cellular metabolism (in that the membrane forms an enclosure that brings molecules in close enough contact that biochemical metabolic reactions can occur at all). The cell is therefore a recursive, autopoietic, or self-making system, a kind of organization that continues to recreate itself.

All of this is quite congenial to and compatible with the contingentist views presented herein, with a few important caveats and clarifications. First, though autopoietic, second-order cybernetics, and systems theories all articulate a close interdependence between organism and environment, where (particularly in autopoietic theory) living beings "bring forth a world by living" (Maturana and Varela, *The Tree of Knowledge*), the question of "ground" or where the world is brought forth *from* remains controversial and unresolved. I read in much of this literature an ambiguity (or perhaps a silence) regarding the external environment, where an intrinsically existent physicalist external environment seems to be posited, albeit one that is inchoate or unformed prior to the cognitive work of organisms whereby it is "brought into being" meaning something like "given form from formlessness." This assumption has been discussed and argued against in Chapter 2 in the section on assumption of nonimpingement. To the extent that autopoeitic, second-order cybernetic, and systems theoretic views do not (explicitly or implicitly) take on this assumption, they are compatible with contingentism. An example of an account in this literature that does not take on the assumption of intrinsically existent environments, no matter how subtle, is one of Francisco Varela's early works, *Principles of Biological Autonomy* (New York: Elsevier North Holland, 1979). Second, thoroughly nonessentialist positions, like the phenomenological approach presented in Evan Thompson's *Mind in Life: Biology, Phenomenology, and the Sciences of Mind* (Cambridge, MA: Belknap Press, 2010), may be read and have sometimes indeed been interpreted as *idealism*, where mind is everything, is essential, and is fundamental. This work can also be read as (and I believe it is) solidly contingentist and as therefore emphatically *non*-idealist, as explained in the previous section, "Assumption of a Ground."

Why do these careful distinctions matter? To the extent that autopoietic, second-order cybernetic, and systems theoretic perspectives collapse, explicitly or implicitly, back onto subtle physicalist or idealist views, they cannot offer a coherent, self-consistent, and precise alternative to both. The absence of this alternative—or, even the readings of these literatures as providing no real alternative—are part of what, I believe, help open the door to the returns to essentialisms (albeit subtle ones) appearing in the forms of speculative realism and object-oriented ontologies; see the notes of the section "Assumption of Knowledge as Limited" in Chapter 5 for more on this topic. To clearly articulate recursion and thorough interdependence without hedges or half-measures that potentially open the door to essentialism is, I believe, critical to the future of both systems theories and of ontology.

The specific use of recursion employed in this section refers to matter, organisms, and experiences as mutually entailing, mutually constitutive, each the "enabling condition" of the others, and therefore *recursive* in the sense of having a looping topology, characteristic of *systems*. "A system" is distinguishable from "a network" by just this topology: a network can be open, whereas a system must have an element of feedback, recursion, closure (B. Clarke, personal communication, 2014). The system outlined here is matter-organism-experience. Interestingly, however, the distinction between network (as an open configuration) and system (as a closed one) converge here, if we consider all of "what is" to be the network or system. For if there is nothing *but* the network, then what does "open" even mean? Such a network is equally open and closed: open because there is nothing it is closed off from, closed because there is nothing it is open to. The same might be said of a system: a closed configuration—but again, what can it be closed off from? This formulation can be compared and contrasted with the following formulation of autopoietic theory: "to maintain their autopoiesis, (self-referential) systems must remain operationally (or organizationally) closed to information from the environment. On that basis, they can construct their interactions with their environment *as* information" (Clarke and Hansen, 2009). In conclusion, though contingentism articulates reality as loopy, recursive, and arising from fully mutually enabling conditions, it is important to be clear that such a system is not precisely closed (to information, to an environment, to an external world, to a reality beyond itself) just because contingentism posits no such intrinsically existing "outside" to the network/system, that is, to contingently existent, mutually enabling phenomena themselves. It *all* depends, without a remainder that can be placed *outside* the "it."

5. WHAT DOES ORDER DEPEND ON? PATTERNS, GAPS, AND THE KNOWN WORLD

1. J. W. Hayward and F. J. Varela, *Gentle Bridges: Conversations with the Dalai Lama on the Sciences of Mind* (San Francisco: Shambhala Publications, 1992).

2. Again, see J. Cooper, *Cognitive Dissonance: 50 Years of a Classic Theory* (London: Sage Publications, 2007), and B. Gawronski and F. Strack, eds., *Cognitive Consistency: A Fundamental Principle in Social Cognition* (New York: Guilford Press, 2012) for reviews of current issues in cognitive dissonance and social psychology.

3. Physical substance here is used in the essentialist sense that assumes some fundamental separation between the external world and perception of that world.

4. In another lucid dream I had, I asked someone to go outside with me. She refused and said, "Just because this is your dream, you think you can do whatever you want." This definitely had me questioning my agency in my own dreams!

5. This is related to the idea presented in Chapter 3 that sensing is itself a type of assimilation. In this view, what is sensed as external becomes a part of the self.

6. "Permitted" here means not given explicit sanction by a social group, of course, but rather, the experience that one can now be surprised. We "permit ourselves" to be surprised.

7. See Barrett (2012) for lucid and vividly illustrated accounts of animal memory as relying not on "storage" of past events but on reliable coordination of body and environment in the present.

8. Philosopher Quentin Meillassoux calls his philosophy "speculative realism," a relatively new and popular philosophical position that begins with a sense of frustration that the "ancestral" preobserver universe is (he claims) effectively off limits to those who see objects as being constituted themselves by observers, a view he calls "correlationism." Meillassoux calls this limit upon the capacity to claim the *intrinsic existence* of preobserver objects as "finitude" and sees his own position as offering a future for philosophy "after finitude"—the title of his book. I take issue with this premise on the grounds that the idea of "finitude" or "limits" is itself a kind of experience and only arises when contrasted with the (unfounded) belief that knowledge should *not* depend upon the very things that make it possible, from bodies to language to objects—all contingently existent.

Meillassoux criticizes the view that objects and organisms (or, in his language, being and thinking) constitute one another as a position that is woefully constraining, and leaves theorists shut off from the absolute. He uses the metaphor of the vicious circle to describe the situation in which "what is" ("the absolute," in his language) cannot be considered independently of thought. He also uses the metaphor of Ptolemaic cosmology (where the Sun was thought to revolve around the Earth) to express what he sees as the subject-centeredness (self-centeredness, human-centeredness, life-centeredness, or thought- centeredness) of any thought system that does not consider the existence of things independently of organisms that relate to them as things.

Can contingentism express vastness instead of the sense of imprisonment and constraint that Meillassoux emphasizes? Can it express not a pessimism about the possibility of accessing absolute knowledge but rather a positive view of knowledge and its relation with the absolute? And can any view coherently preserve both the *vastness* of what is *and* its *accessibility* to living beings? I answer each of these questions in the affirmative.

What if the real world—called by other thinkers things like the Absolute, the Truth, or what *is*—is not a vista that can only be viewed through a tiny chink in an otherwise opaque door? What if the real world is utterly mundane, too close and too ordinary to be noticed, an open secret? If so, the real world is accessible not only through the very specific methods and practices of rationalist philosophy and mathematics as Meillassoux suggests, but also through ordinary living. Organisms—including bees, cats, and biologists—are then not primordially ignorant, but are intimate with the real. Contingentism and the "network" of dependencies it describes then become not a web in which we are constrained and caught, but an expression of the vastness of lives and of everyday knowledge.

Again, a sense of impoverishment, the intuition that *we do not have everything we need* in order to access *genuine* knowledge and the *real* world, may be part of what gives rise to returns to essentialism. Contingentism offers a workable alternative to this sense of impoverishment.

REFERENCES

Adler, J., and W-W. Tso. (1974). "Decision"-making in bacteria: chemotactic response of *Escherichia coli* to conflicting stimuli. *Science* 184 (4143): 1292–1294.

Ahlquist, R. P. (1948). A study of the adrenotrophic receptors. *American Journal of Physiology* 155: 586–600.

Ashcroft, S. J. H., J. R. Crossley, and P. C. C. Crossley. (1974). Effect of *N*-acylglucosamines on the biosynthesis and secretion of insulin in the rat. *Biochemical Journal* 154: 701–707.

Barrett, L. (2012). *Beyond the Brain: How Body and Environment Shape Animal and Human Minds*. Princeton: Princeton University Press.

Batko, A., and J. Jakubiec. (1989). *Gonium dispersum*, a new species of *Gonium* from Poland. *Archiv für Hydrobiologie Supplement* 82 (1): 39–48.

Birch, C. (1999). *Biology and the Riddle of Life*. Sydney: University of New South Wales Press.

Botvinick, M., and J. Cohen. (February 1998). Rubber hands "feel" touch that eyes see. *Nature* 391 (6669): 756.

Brook, A., and P. Raymont. (2014). The unity of consciousness. In *The Stanford Encyclopedia of Philosophy* (Spring 2014 Edition), ed. Edward N. Zalta. Retrieved on May 2, 2014, from http://plato.stanford.edu/archives/spr2014/entries/consciousness-unity.

Burbulys, D., K. A. Trach, and J. A. Hoch. (1991). Initiation of sporulation in *Bacillus subtilis* is controlled by a multicomponent phosphorelay. *Cell* 64: 545–552.

Butler, J. (2006). *Precarious Life: The Powers of Mourning and Violence*. London: Verso.

Byrne, R. (2006). *The Secret*. New York: Atria Books.

Clark, A. J. (1933). *The Mode of Action of Drugs on Cells*. London: Edward Arnold & Co.

Clarke, B., and M. Hansen. (2009). Neocybernetic emergence: Returning the posthuman. *Cybernetics and Human Knowing* 16 (1–2): 83–99.

Cooper, J. (2007). *Cognitive Dissonance: 50 Years of a Classic Theory*. London: Sage.

Coventry, A. (2006). *Hume's Theory of Causation: A Quasi-Realist Interpretation*. London: Continuum.

Davis, A., J. Izatt, and F. Rothenberg. (2009). Quantitative measurement of blood flow dynamics in embryonic vasculature using spectral Doppler velocimetry. *Anatomical Record* 292 (3): 311–319.

Deacon, T. W. (1997). *The Symbolic Species: The Co-evolution of Language and the Brain.* New York: Norton.

de Waal, F. (2006). *Primates and Philosophers: How Morality Evolved.* Princeton: Princeton University Press.

Di Paolo, E., and E. Thompson. (2014). The enactive approach. In *The Routledge Book of Embodied Cognition,* ed. Lawrence Shapiro. New York: Routledge.

Douglas, R. H., C. W. Mullineaux, & J. C Partridge. (2000). Long-wave sensitivity in deep-sea stomiid dragonfish with far-red bioluminescence: evidence for a dietary origin of the chlorophyll-derived retinal photosensitizer of Malacosteus niger. *Philosophical Transactions of the Royal Society B: Biological Sciences, 355*(1401), 1269–1272.

Douglas, R. H., J. C. Partridge, K. Dulai, D. Hunt, C. W. Mullineaux, A. Y. Tauber, and P. H. Hynninen. Dragonfish see using chlorophyll. *Nature 393* (4 June 1998): 423–424.

Falke, J. J. R., R. B. Bass, S. L. Butler, S. A. Chervitz, and M. A. Danielson. (1997). The two-component signaling pathway of bacterial chemotaxis: A molecular view of signal transduction by receptors, kinases, and adaptation enzymes. *Annual Reviews in Cellular and Developmental Biology* 13: 457–512.

Feynman, R. (1964). *The Feynman Lectures on Physics: Volume 1.* New York: Addison-Wesley.

———. (1989). *QED: A Strange Theory of Light and Matter.* Princeton: Princeton University Press.

Fleck, L. (1979). *Genesis and Development of a Scientific Fact.* Chicago: University of Chicago Press.

Garfield, J. L. (1990). Epoché and Śūnyatā: Skepticism East and West. *Philosophy East and West* 40: 285–307.

———. (1995). *The Fundamental Wisdom of the Middle Way: Nagarjuna's Mulamadhyamaka-karika.* Oxford: Oxford University Press.

———. (2015). Just Another Word for Nothing Left to Lose: Freedom, Agency and Ethics for Mādhyamikas. In press. Pre-print retrieved February 22, 2015, from http://www.smith.edu/philosophy/faculty_garfield.php.

Gawronski, B., and F. Strack, eds. (2012). *Cognitive Consistency: A Fundamental Principle in Social Cognition.* New York: Guilford Press.

Gibson, J. J. (1979). *The Ecological Approach to Visual Perception.* Boston: Houghton Mifflin

Gomperts, B. D., I. M. Kramer, and P. E. Tatham. (2002). *Signal Transduction.* Orlando, FL: Academic Press.

Griffiths, David. (2008). *Introduction to Elementary Particles.* 2nd edition. Weinheim, Germany: Wiley-VCH Verlag GmbH & Co.

Hammarsten, O, and J. A. Mandel. (1904). *A Text-book of Physiological Chemistry.* 4th edition. New York: John Wiley & Sons.

Hayward, J. W., and F. J. Varela. (1992). *Gentle Bridges: Conversations with the Dalai Lama on the Sciences of Mind.* San Francisco: Shambhala.

Heider, F. & Simmel, M. (1944). An experimental study of apparent behavior. *American Journal of Psychology* 57: 243–249.

Hellingwerf, K. J., W. M. Crielaard, M. J. Teixeira deMattos, W. D. Hoff, R. Kort, D. T. Verhamme, and C. Avignone-Rossa. (1998). Current topics in signal transduction in bacteria. *Antonie van Leeuwenhoek* 74: 211–227.

Hildebrand, E. (1977). What does *Halobacterium* tell us about photoreception? *Biophysical Structural Mechanics* 3: 69–77.

Hofstadter, D. R. (1979). *Gödel, Escher, Bach: An Eternal Golden Braid.* New York: Basic Books.

———. (2007). *I Am a Strange Loop.* New York: Basic Books.

Hume, D. (1888). *Treatise of Human Nature.* New York: Macmillan and Co.

Jeffery, D. (2009). The concept of energy in physics. Retrieved October 7, 2009 from http://physics.nhn.ou.edu/~jeffery/course/c_energy/energy1/lec001.html

Keller, E. F. (1992). *Secrets of Life, Secrets of Death.* New York: Routledge.

Kirk, David L. (2005). A twelve-step program for evolving multicellularity and a division of labor. *BioEssays* 27: 299–310.

Langley, J. N. (1905). On the reaction of cells and of nerve-endings to certain poisons, chiefly as regards the reaction of striated muscle to nicotine and to curare. *Journal of Physiology* 33: 374–413.

Latour, B. (1993). *The Pasteurization of France.* Translated by Alan Sheridan and John Law. Cambridge, MA: Harvard University Press.

———. (2013). *An Inquiry into Modes of Existence: An Anthropology of the Moderns.* Translated by Catherine Porter. Cambridge, MA: Harvard University Press.

Lloyd, G. E. R. (2009). *Cognitive Variations: Reflections on the Unity and Diversity of the Human Mind.* Oxford: Oxford University Press.

Maturana, H. R., and F. Varela. (1980). *Autopoiesis and Cognition: The Realization of the Living.* Dordrecht: Reidel.

———. (1992). *The Tree of Knowledge: The Biological Roots of Human Understanding.* Boston: Shambhala Publications.

Meillassoux, Q. (2010). *After Finitude: Essays on the Necessity of Contingency.* London: Continuum.

Monod, J. (1971). *Chance and Necessity: An Essay on the Natural Philosophy of Modern Biology.* New York: Knopf.

Noë, A. (2009). *Out of Our Heads: Why You Are Not Your Brain and Other Lessons from the Biology of Consciousness.* New York: Hill and Wang.

Odling-Smee, F. J., K. N. Laland, and M. F. Feldman. (2003). *Niche Construction: The Neglected Process in Evolution.* Princeton: Princeton University Press.

Oyama, S. (2000). *The Ontogeny of Information.* 2nd edition. Durham, NC: Duke University Press.

Pipeleers, D. G., M. Marichal, and W. J. Malaisse. (1973). The stimulus-secretion coupling of glucose-induced insulin release. XIV. Glucose regulation of insular biosynthetic activity. *Endocrinology* 93: 1001–1017.

Rodbell, M. (1994). Signal transduction: Evolution of an idea. Nobel lecture. Retrieved on June 5th, 2009, from http://nobelprize.org/nobel_prizes/medicine/laureates/1994/rodbell-lecture.pdf.

Rørslett, B. (October 24, 2006). Flowers in ultraviolet: Arranged by plant family. Retrieved May 4, 2014, from http://www.naturfotograf.com/index2.html.

Rorty, R. (1979). *Philosophy and the Mirror of Nature.* Princeton: Princeton University Press.

Savage-Rumbaugh, S., S. G. Shanker, and T. J. Taylor. (1998). *Apes, Language, and the Human Mind.* New York: Oxford University Press.

Seager, W., and S. Allen-Hermanson. (2013). Panpsychism. *The Stanford Encyclopedia of Philosophy* (Fall 2013 Edition), ed. Edward N. Zalta. http://plato.stanford.edu/archives/fall2013/entries/panpsychism.

Smith, B. H. (2006). *Scandalous Knowledge: Science, Truth, and the Human.* Durham, NC: Duke University Press.

Thompson, E. (2010). *Mind in Life: Biology, Phenomenology, and the Sciences of Mind.* Cambridge, MA: Belknap Press.

Varela, F. J. (1979). *Principles of Biological Autonomy.* New York: Elsevier North Holland.

Varela, F. J., E. Thompson, and E. Rosch. (1991). *The Embodied Mind: Cognitive Science and Human Experience.* Cambridge, MA: MIT Press.

Wanner, B. L. (1992). Is cross regulation by phosphorylation of two-component response regulator proteins important in bacteria? *Journal of Bacteriology* 174: 2053–2058.

Washington, I., J. Zhou, S. Jockusch, N. J. Turro, K. Nakanishi, and J. R. Sparrow. (2007). Chlorophyll derivatives as visual pigments for super vision in the red. *Photochemical and Photobiological Science* 6 (7): 775–779.

West-Eberhard, M. J. (2003). *Developmental Plasticity and Evolution.* New York: Oxford University Press.

Whitehead, A. N. (1929). *Process and Reality: An Essay in Cosmology.* New York: Macmillan.

Wiener, N. (1948). *Cybernetics: Or, Control and Communication in the Animal and the Machine.* Cambridge, MA: MIT Press.

INDEX

stasis, change and, 59

Stoeckenius, W., 49

strange loop, 82

Straub, Walter, 45

subjects, 4; gap between objects, 97; intrinsic existence of other, 77–79

substance, energy as, 57–60

substantiality, 11

sugar sensing cells, 8–9

surprise, 90–91

TCSTS (two-component signal transduction systems), 42–43

thoughts, essentialism and, 11

transducers: cybernetic theory and, 46; energy and, 54; *versus* receptors, 45

ultraviolet light, 9–10, 107n4

unicellular organisms, as model systems, 49–50

unified object of sense perception, 33–37

The Unity of Consciousness (Cleeremans), 74

unobserved phenomena, existence and, 14–15

validation of knowledge, 33–37

Varela, Francisco, 34–35

Verworn, Max, 44

vision, ultraviolet light, 9–10, 107n4

visual perception, 18–19

visualization, perception and, 84

Wiener, Norbert, *Cybernetics*, 48

will, 87; events and, 88

wonder, 103–104

ACKNOWLEDGMENTS

What Does This Book Depend On?

On the living: on my grandfather tending the hummingbird feeder, and on the ruby-throated rushing that held him rapt; on what holds us together and holds us in thrall, corals in seawater, elephants in packs. On the amoebae in the spore body and on those who die to make the stalk; on cuttlefish, on tardigrades, on every last sequoia.

I went off for days and days, quietly sitting, thinking, and writing, in what appeared to be solitude. But when was I ever alone? Wherever I am, there *you* are—driving carefully to keep me alive, making the atmosphere to keep me alive, raising crops to keep me alive. What would I do or be without you? You're the reason that anything works at all around here. You have my gratitude, always.

On the generosity and intelligence of my colleagues at Duke University Biology Department: on Louise Roth, Amy Schmid, Paul Magwene, Carl Simpson, Ed Venit, Debra Murray, Josh Granek, Omur Kayikci, Keely Dulmage, and Horia Todor. On David McCandlish, Viviane Callier, and Lauren McCall, who clarified, criticized, and carefully encouraged the right questions. On Dan McShea most of all, the very best of advisors.

On rigorous and clarifying conversations with Rick Dilling, Carl Rothfels, Kelly Foyle, Simon Kiss, Jay Garfield, and Geoff Driscoll, traces of which can be found in these pages.

On Srinivas Aravamudan, who generously offered fellowship support.

On Evan Thompson, who gave a careful and encouraging reading, and perceptive suggestions that strengthened the book.

On Susan Oyama, who kept my work in mind, whose work I keep in mind always, who brought the curiosity and lucidity she brings to all things to this project.

On Barbara Herrnstein Smith, most of all—on her vital feedback, generous guidance, and unflagging support at every step.

On my colleagues at University of North Carolina at Chapel Hill. On Beth Shank who supported and encouraged in the last—and therefore most demanding—miles of this marathon. On Arturo Escobar and Michal Osterweil, who created critical interdisciplinary alliances, and who continue to nurture knowledge into something we might call wisdom.

On Jay Garfield, who does the difficult and important work of writing about existence in a lucid and nuanced way, and offered encouragement and guidance as I attempted the same.

On so many teachers. In particular, on Nico Bethel, who made everything strange and new.

On Helen Tartar, who offered the book contract a week before her death, and who I wish could see it today. I hope this piece is worthy of her irreplaceable vision, and that it lives still in these pages. On Tom Lay, who brought this work to completion with consideration for both the book and its author. On thoughtful copy editing by Gregory McNamee. On Bruno Clarke, who invited the book into the fold of this series and offered crucial clarifications that helped it become well integrated into this fine company. On Fordham University Press, which brings books into being with tremendous humanity.

On the wise and kind, my commun(ivers)ity, the brilliant and rigorous thinkers, actors, and teachers who shaped my mind and life in the years of writing *Interdependence*; who taught me the phrase "I am because we are"; and who puzzled and lived its meaning with me: Afiya Carter, Aiden Graham, AJ Vrieze, Alexis Pauline Gumbs, Atiya Hussain, Beth Bruch, Bryan Proffitt, Caitlin Breedlove, Dannette Sharpley, Emily Chavez, Glenys Verhulst, Jenn Vrieze, Jurina Vincent-Lee, Kai Barrow, Keagha Carscallen, Laurin Penland, Lynne Walter, Manju Rajendran, Marjorie Scheer, Michelle O' Brien, Mikel Barton, Monica Leonardo, Nia Wilson, Nikki Brown, Noah Blose, Paulina Hernandez, Pavithra Vasudevan, Rachael Derello, Russell Herman, Sam Hummel, Sammy Truong, Sendolo Diaminah, Serena Sebring, Shirlette Ammons, Tema Okun, Theo Luebke, Tim Stallman, Tony Macias, and Yolanda Carrington.

On my generous, loving family: Sharmas, Nagars, Bhardwajs, Joshis, Kakrias, Maxwells, and Partins all, near and far, living and beyond.

On Colin Maxwell, who illustrated the line figures; who engaged every new idea thoughtfully and energetically; who brought philosophical acumen, good advice, and a great many cups of coffee; and who I am lucky to call *jeevansaathi*.

On Mom and Dad, who make all things possible. On Sachi and Neel-Gopal—here's a nerdy analogy you'll appreciate: in early development, the chick heart is a pumping tube that isn't strictly necessary for oxygen circulation, but helps create the body's form by its pumping. You are the early heart that made me, and this book.

I believe that books find their people. I have believed this because of how countless books have come into my life seemingly at just the time when I have needed them. How does this consistently come to pass? It has been a mystery to me, but now I think I know: If a book seems to find you, it's because it has been delivered to you by many hands.